Chatter

Chatter

THE VOICE IN OUR HEAD,
WHY IT MATTERS, *and*
HOW TO HARNESS IT

Ethan Kross

CROWN

NEW YORK

Published in the United States by Crown, an imprint of Random House, a division of Penguin Random House LLC, New York.

CROWN and the Crown colophon are registered trademarks of Penguin Random House LLC.

Library of Congress Cataloging-in-Publication Data
Names: Kross, Ethan, author.
Title: Chatter / Ethan Kross.
Description: First edition. | New York : Crown, [2021] | Includes bibliographical references and index.
Identifiers: LCCN 2020025201 (print) | LCCN 2020025202 (ebook) | ISBN 9780525575238 (hardcover) | ISBN 9780593238752 (international edition) | ISBN 9780525575252 (ebook)
Subjects: LCSH: Self-talk. | Thought and thinking. | Communication—Psychological aspects.
Classification: LCC BF697.5.S47 K76 2021 (print) | LCC BF697.5.S47 (ebook) | DDC 158.1—dc23
LC record available at https://lccn.loc.gov/2020025201
LC ebook record available at https://lccn.loc.gov/2020025202

Printed in the United States of America on acid-free paper

randomhousebooks.com

4689753

Book design by Elizabeth Rendfleisch

To Dad, for teaching me to go inside,
and
Lara, Maya, and Dani, my ultimate chatter antidotes

The biggest challenge, I think, is always maintaining your moral compass. Those are the conversations I'm having internally. I'm measuring my actions against that inner voice that for me at least is audible, is active, it tells me where I think I'm on track and where I think I'm off track.

—BARACK OBAMA

The voice in my head is an asshole.

—DAN HARRIS

Contents

Introduction

I stood in the darkness of my living room, my knuckles white, my fingers tense around the sticky rubber handle of my Little League baseball bat, staring out the window into the night, trying desperately to protect my wife and newborn daughter from a madman I had never met. Any self-awareness about how this looked, or about what I might actually do if the madman appeared, had been washed away by the fear I was experiencing. The thoughts racing through my head kept repeating the same thing.

It's all my fault, I said to myself. *I have a healthy, adorable new baby and wife upstairs who love me. I've put them both at risk. What have I done? How am I going to fix this?* These thoughts were like a horrible carnival ride I couldn't get off.

So there I was, trapped—not just in my dark living room, but also in the nightmare of my own mind. Me, a scientist who directs a laboratory that specializes in the study of *self-control,* an expert on how to tame unrelenting negative thought spirals, staring out the window at three in the morning with a tiny baseball bat in my hands, tortured not by the boogeyman who sent me a deranged letter but by the boogeyman inside my head.

How did I get here?

The Letter and the Chatter

That day began like any other day.

I woke up early, got dressed, helped feed my daughter, changed her diaper, and quickly downed breakfast. Then I kissed my wife and headed out the door to drive to my office on the University of Michigan's campus. It was a cold but tranquil, sunny day in the spring of 2011, a day that seemed to promise equally tranquil, sunny thoughts.

When I arrived at East Hall, the mammoth brick-covered building that houses the University of Michigan's storied Psychology Department, I found something unusual in my mailbox. Sitting atop the stack of science journals that had been accumulating was an envelope hand addressed to me. Curious about what was inside—it was rare that I received hand-addressed mail at work—I opened the letter and began reading it as I walked toward my office. That's when, before I even realized how hot I was, I felt a rush of sweat slide down my neck.

The letter was a threat. The first one I had ever received.

The previous week I had appeared briefly on *CBS Evening News* to talk about a neuroscience study that my colleagues and I had just published demonstrating that the links between physical and emotional pain were more similar than previous research had suggested. In fact, the brain registered emotional and physical pain in remarkably similar ways. Heartbreak, it turned out, was a physical reality.

My colleagues and I had been excited about the results yet didn't expect them to generate more than a handful of calls from science journalists looking to file a brief story. Much to our surprise, the findings went viral. One minute I was lecturing to undergraduates on the psychology of love, and the next I was receiving a crash course in media training in a television studio on campus. I managed to get through the interview without tripping on my words too many times, and a few hours later the segment on our work aired—a scientist's fifteen minutes of fame, which in fact amounted to about ninety seconds.

What exactly our research had done to offend the letter writer wasn't clear, but the violent drawings, hateful slurs, and disturbing messages that the text contained left little to my imagination about the person's feelings toward me while at the same time leaving much to my imagination about what form such malice could take. To make matters worse, the letter didn't come from a distant locale. A quick Google search of its postmark revealed that it was sent from just a dozen miles away. My thoughts started spinning uncontrollably. In a cruel twist of fate, I was now the one experiencing emotional pain so intense it felt physical.

Later that day, after several conversations with university administrators, I found myself sitting in the local police station, anxiously awaiting my turn to speak to the officer in charge. Although the policeman I eventually shared my story with was kind, he wasn't particularly reassuring. He offered three pieces of advice: Call the phone company and make sure my home telephone number wasn't listed, keep an eye out for suspicious people hanging around my office, and—my personal favorite—drive home from work a different way each day to ensure that no one learned my routine. That was it. They were not deploying a special task force. I was on my own. It was not exactly the comforting response I had hoped to hear.

As I took a long, circuitous route home that day through Ann Arbor's tree-lined streets, I tried to come up with a solution for how to deal with the situation. I thought to myself, *Let's go over the facts. Do I need to worry? What do I need to do?*

According to the police officer, and several other people I had shared my story with, there were clear ways I could answer these questions. *No, you don't need to stress out over this. These things happen. There isn't anything else you can do. It's okay to be afraid. Just relax. Public figures receive empty threats all the time and nothing happens. This will blow over.*

But that wasn't the conversation I had with myself. Instead, the despairing stream of thoughts running through my head amplified itself in an endless loop. *What have I done?* my inner voice shouted, before switching into my inner frenzy maker. *Should I call the alarm company? Should I get a gun? Should we move? How quickly can I find a new job?*

A version of this conversation repeated itself again and again in my mind over the next two days, and I was a nervous wreck as a result. I had no appetite, and I talked endlessly (and unproductively) about the threatening letter with my wife to the point that tension between us began to grow. I startled violently each time I heard the faintest peep escape from my daughter's nursery, instantly assuming that the worst fate was upon her rather than a more obvious explanation—a creaky crib, a gassy baby.

And I paced.

For two nights, while my wife and daughter slept peacefully in their beds, I stood watch downstairs in my pajamas with my Little League baseball bat in my hands, peeking out the living room window to make sure no one was approaching, with no plan for what I would do if I actually found someone lurking outside.

At my most embarrassing, when my anxiety peaked on the second night, I sat down in front of my computer and considered performing a Google search with the key words "bodyguards for academics"—absurd in hindsight but urgent and logical at the time.

Going Inside

I am an experimental psychologist and neuroscientist. I study the science of introspection at the Emotion & Self Control Laboratory, a lab I founded and direct at the University of Michigan. We do research on the silent conversations people have with themselves, which powerfully influence how we live our lives. I've spent my entire professional career researching these conversations—what they are, why we have them, and how they can be harnessed to make people happier, healthier, and more productive.

My colleagues and I like to think of ourselves as mind mechanics. We bring people into our lab to participate in elaborate experiments, and we also study them "in the wild" of daily human experience. We use tools from psychology and other disciplines—fields as diverse as medicine, philosophy, biology, and computer science—to answer vexing questions like: Why are some people able to benefit from focusing inward to understand their feelings, while other people crumble when they engage in the exact same behavior? How can people reason wisely under toxic stress? Are there right and wrong ways to talk to yourself? How can we communicate with people we care about without stoking their negative thoughts and emotions or increasing our own? Do the countless "voices" of others we encounter on social media affect the voices in our minds? By rigorously examining these questions, we've made numerous surprising discoveries.

We've learned how specific things we say and do can improve our inner conversations. We've learned how to pick the locks of the "magical" back doors of the brain—how certain ways of employing placebos, lucky charms, and rituals can make us more resilient. We've learned which pictures to place on our desks to help us recover from emotional injuries (hint: photos of Mother Nature can be comforting just like those of our own mothers), why clutching a stuffed animal can help with existential despair, how and how not to talk with your partner after a hard day, what you're likely doing wrong when you log on to social media, and where you should go when you take walks to deal with the problems you face.

My interest in how the conversations we have with ourselves influence our emotions began long before I considered a career in science. It began before I really understood what feelings were. My fascination with the rich, fragile, and ever-shifting world we carry around between our ears dates back to the first psychology lab I ever set foot in: the household where I grew up.

I was raised in the working-class Brooklyn neighborhood of Canarsie to a father who taught me about the importance of self-reflection from an oddly early age. When I suspect the parents of most other three-year-olds were teaching their kids to brush their teeth regularly and treat other people kindly, my dad had other priorities. In his typically unconventional style, he was more concerned with my inner choices than anything else, always encouraging me to "go inside" if I had a problem. He liked to tell me, "Ask yourself *the* question." The exact question he was referring to eluded me, though on some level I understood what he was pushing me to do: *Look inside yourself for answers.*

In many ways, my dad was a walking contradiction. When he wasn't flipping off other drivers on noisy, traffic-choked New York streets or cheering on the Yankees in front of the television at home, I could find him meditating in his bedroom (usually with a cigarette dangling beneath his bushy mustache) or reading the Bhagavad Gita. But as I grew up and encountered situations more complex than deciding whether to eat a forbidden cookie or refusing to clean my room, his advice took on more weight. Should I ask my high school crush out? (I did; she said no.) Should I confront my friend after witnessing him steal someone's wallet? Where should I go to college? I prided myself on my coolheaded thinking, and my reliance on "going inside" to help me make the right decision rarely faltered (and one day one of my crushes would say yes; I married her).

Perhaps unsurprisingly, when I went off to college, my discovery of the field of psychology felt preordained. I had found my calling. It explored the things my dad and I had spent my youth talking about when we weren't talking about the Yankees; it seemed to both explain my childhood and show me a pathway into adulthood. Psychology also gave me a new vocabulary. In my college classes, I learned, among lots of other things, that what my father had been circling around during all those years of his Zen parenting, which my markedly not-eccentric mother had put up with, was the idea of introspection.

In the most basic sense, introspection simply means actively paying attention to one's own thoughts and feelings. The ability to do this is what allows us to imagine, remember, reflect, and then use these reveries to problem solve, innovate, and create. Many scientists, including myself, see this as one of *the* central evolutionary advances that distinguishes human beings from other species.

All along, then, my father's rationale was that cultivating the skill of introspection would help me through whatever challenging situations I encountered. Deliberate self-reflection would lead to wise, beneficial choices and by extension to positive emotions. In other words, "going inside" was the route to a resilient, fulfilling life. This made perfect sense. Except that, as I would soon learn, for many people it was completely wrong.

In recent years, a robust body of new research has demonstrated that when we experience distress, engaging in introspection often does significantly more harm than good. It undermines our performance at work, interferes with our ability to make good decisions, and negatively influences our relationships. It can also promote violence and aggression, contribute to a range of mental disorders, and enhance our risk of becoming physically ill. Using the mind to engage with our thoughts and feelings in the wrong ways can lead professional athletes to lose the skills they've spent their careers perfecting. It can cause otherwise rational, caring people to make less logical and even less moral decisions. It can lead friends to flee from you in both the real world and the social media world. It can turn romantic relationships from safe havens into battlegrounds. It can even contribute to us aging faster, both in how we look on the outside and in how our DNA is configured internally. In short, our thoughts too often *don't* save us from our thoughts. Instead, they give rise to something insidious.

Chatter.

Chatter consists of the cyclical negative thoughts and emotions that turn our singular capacity for introspection into a curse rather than a blessing. It puts our performance, decision making, relationships, happiness, and health in jeopardy. We think about that screwup at work or misunderstanding with a loved one and end up flooded by how bad we feel. Then we think about it again. And again. We introspect hoping to tap into our inner coach but find our inner critic instead.

The question, of course, is *why*. Why do people's attempts to "go inside" and think when they experience distress at times succeed and at other times fail? And just as important, once we find our introspective abilities running off course, what can we do to steer them back on track? I've spent my career examining these questions. I've learned that the answers hinge on changing the nature of one of the most important conversations of conscious life: the ones we have with ourselves.

Our Default State

A widespread cultural mantra of the twenty-first century is the exhortation to *live in the present*. I appreciate the wisdom of this maxim. Instead of succumbing to the pain of the past or anxiety about the future, it advises, we should concentrate on connecting with others and oneself right now. And yet, as a scientist who studies the human mind, I can't help but note how this well-intentioned message runs counter to our biology. Humans weren't made to hold fast to the present all the time. That's just not what our brains evolved to do.

In recent years, cutting-edge methods that examine how the brain processes information and allow us to monitor behavior in real time have unlocked the hidden mechanics of the human mind. In doing so, they have uncovered something remarkable about our species: We spend one-third to one-half of our waking life *not* living in the present.

As naturally as we breathe, we "decouple" from the here and now, our brains transporting us to past events, imagined scenarios, and other internal musings. This tendency is so fundamental it has a name: our "default state." It is the activity our brain automatically reverts to when not otherwise engaged, and often even when we are otherwise engaged. You've no doubt noticed your own mind wander, as if of its own volition, when you were supposed to be focusing on a task. We are perpetually slipping away from the present into the parallel, nonlinear world of our minds, involuntarily sucked back "inside" on a minute-to-minute basis. In light of this, the expression "the life of the mind" takes on new and added meaning: Much of our life *is* the mind. So, what often happens when we slip away?

We talk to ourselves.

And we listen to what we say.

Humanity has grappled with this phenomenon since the dawn of civilization. Early Christian mystics were thoroughly annoyed by the voice in their head always intruding on their silent contemplation. Some even considered these voices demonic. Around the same time, in the East, Chinese Buddhists theorized about the turbulent mental weather that could cloud one's emotional landscape. They called it "deluded thought." And yet many of these very same ancient cultures believed that their inner voice was a source of wisdom, a belief that undergirds several millennia-old practices like silent prayer and meditation (my dad's personal philosophy). The fact that multiple spiritual traditions have both feared our inner voice and noted its value speaks to the ambivalent attitudes to our internal conversations that still persist today.

When we talk about the inner voice, people naturally wonder about its pathological aspects. I often begin presentations by asking audience members if they talk to themselves in their heads. Invariably, many people look relieved to see other hands shoot up alongside theirs. Unfortunately, normal voices that we hear in our heads (belonging, for example, to ourselves, family, or colleagues) can sometimes devolve into abnormal voices characteristic of mental illness. In such cases, the person doesn't believe that the voice issues from their own mind but thinks it comes from another entity (hostile people, aliens, and the government, to name a few common auditory hallucinations). Importantly, when we talk about the inner voice, the difference between mental illness and wellness is a question not of dichotomy—pathological versus healthy—but of culture and degree. One quirk of the human brain is that roughly one in ten people hear voices and attribute them to external factors. We are still trying to understand why this happens.

The bottom line is that we all have a voice in our head in some shape or form. The flow of words is so inextricable from our inner lives that it persists even in the face of vocal impairments. Some people who stutter, for example, report talking more fluently in their minds than they do out loud. Deaf people who use sign language talk to themselves too, though they have their own form of inner language. It involves silently signing to themselves, similar to how people who can hear use words to talk to themselves privately. The inner voice is a basic feature of the mind.

If you've ever silently repeated a phone number to memorize it, replayed a conversation imagining what you should have said, or verbally coached yourself through a problem or skill, then you've employed your inner voice. Most people rely on and benefit from theirs every day. And when they disconnect from the present, it's often to converse with that voice or hear what it has to say—and it can have a *lot* to say.

Our verbal stream of thought is so industrious that according to one study we internally talk to ourselves at a rate equivalent to speaking four thousand words per minute out loud. To put this in perspective, consider that contemporary American presidents' State of the Union speeches normally run around six thousand words and last over an hour. Our brains pack nearly the same verbiage into a mere sixty seconds. This means that if we're awake for sixteen hours on any given day, as most of us are, and our inner voice is active about half of that time, we can theoretically be treated to about 320 State of the Union addresses each day. The voice in your head is a very fast talker.

Although the inner voice functions well much of the time, it often leads to chatter precisely when we need it most—when our stress is up, the stakes are high, and we encounter difficult emotions that call for the utmost poise. Sometimes this chatter takes the form of a rambling soliloquy; sometimes it's a dialogue we have with ourselves. Sometimes it's a compulsive rehashing of past events (*rumination*); sometimes it's an angst-ridden imagining of future events (*worry*). Sometimes it's a free-associative pinballing between negative feelings and ideas. Sometimes it's a fixation on one specific unpleasant feeling or notion. However it manifests itself, when the inner voice runs amok and chatter takes the mental microphone, our mind not only torments but paralyzes us. It can also lead us to do things that sabotage us.

Which is how you find yourself peeking out the window of

your living room late at night holding a comically small baseball bat.

The Puzzle

One of the most crucial insights I've had during my career is
that the instruments necessary for reducing chatter and harnessing our inner voice aren't something we need to go looking
for. They are often hidden in plain sight, waiting for us to put
them to work. They are present in our mental habits, quirky
behaviors, and daily routines, as well as in the people, organizations, and environments we interact with. In this book, I will
lay bare these tools and explain not only how they work but
how they fit together to form a toolbox that evolution crafted
to help us manage the conversations we have with ourselves.

In the chapters ahead, I will bring the lab to you while also
telling stories about people who combat their chatter. You'll
learn about the mental lives of a former NSA agent, Fred Rogers, Malala Yousafzai, LeBron James, and an indigenous South
Pacific tribe called the Trobrianders, as well as many people just
like you and me. But to begin this book, we will first look at
what the inner voice really is, along with all the marvelous
things it does for us. Then I will take us into the dark side of
the conversations we have with ourselves and the truly frightening extent to which chatter can harm our bodies, damage
our social lives, and derail our careers. This inescapable tension
of the inner voice as both a helpful superpower and destructive
kryptonite that hurts us is what I think of as the great puzzle of
the human mind. How can the voice that serves as our best
coach also be our worst critic? The rest of the chapters will

describe scientific techniques that can reduce our chatter—techniques that are rapidly helping us solve the puzzle of our own minds.

The key to beating chatter isn't to stop talking to yourself. The challenge is to figure out how to do so more effectively. Fortunately, both your mind and the world around you are exquisitely designed to help you do precisely that. But before we get into how to control the voice in our head, we need to answer a more basic question.

Why do we have one in the first place?

Chatter

CHAPTER ONE

Why We Talk to Ourselves

*T*he sidewalks of New York City are superhighways of anonymity. During the day, millions of intent pedestrians stride along the pavement, their faces like masks that betray nothing. The same expressions pervade the parallel world beneath the streets—the subway. People read, look at their phones, and stare off into the great invisible nowhere, their faces disconnected from whatever is going on in their minds.

Of course, the unreadable faces of eight million New Yorkers belie the teeming world on the other side of that blank wall they've learned to put up: a hidden "thoughtscape" of rich and active internal conversations, frequently awash with chatter. After all, the inhabitants of New York are nearly as famous for their neuroses as they are for their gruffness. (As a native, I say this with love.) Imagine, then, what we might learn if we could burrow past their masks to eavesdrop on their inner voices. As it happens, that is exactly what the British anthropologist Andrew Irving did over the course of fourteen months beginning in 2010—listened in on the minds of just over a hundred New Yorkers.

While Irving hoped to gain a glimpse into the raw verbal life of the human mind—or rather an audio sample of it—the origin of his study actually had to do with his interest in how we deal with the awareness of death. A professor at the University of Manchester, he had done earlier fieldwork in Africa analyzing the vocalized inner monologues of people diagnosed with HIV/AIDS. Unsurprisingly, their thoughts roiled with the anxiety, uncertainty, and emotional pain produced by their diagnoses.

Now Irving wanted to compare these findings with a group of people who surely had their woes but weren't necessarily in aggrieved states to begin with. To carry this out, he simply (and bravely!) approached New Yorkers on the street and in parks and cafés, explained his study, and asked if they would be willing to speak their thoughts aloud into a recording device while he filmed them at a distance.

Some days, a handful of people said yes; other days, only one. It was to be expected that most New Yorkers would be too busy or skeptical to agree. Eventually, Irving gathered his one hundred "streams of internally represented speech," as he described them, in recordings ranging from fifteen minutes to an hour and a half. The recordings obviously don't provide an all-access backstage pass to the mind, because an element of performance might have come into play for some participants. Even so, they offer an uncommonly candid window into the conversations people have with themselves as they navigate their daily lives.

As was only natural, prosaic concerns occupied space in the minds of everyone in Irving's study. Many people commented on what they observed on the streets—other pedestrians, drivers, and traffic, for example—as well as on things they needed to do. But existing alongside these unremarkable musings were monologues negotiating a host of personal wounds, distresses, and worries. The narrations often landed on negative content with utterly no transition, like a gaping pothole appearing suddenly on the unspooling road of thought. Take, for example, a woman in Irving's study named Meredith whose inner conversation pivoted sharply from everyday concerns to matters of literal life and death.

"I wonder if there's a Staples around here," Meredith said, before shifting, like an abrupt lane change, to a friend's recent cancer diagnosis. "You know, I thought she was going to tell me that her cat died." She crossed the street, then said, "I was prepared to cry about her cat, and then I'm trying not to cry about her. I mean New York without Joan is just . . . I can't even imagine it." She started crying. "She'll probably be fine, though. I love that line about having a 20 percent chance of being cured. And how a friend of hers said, 'Would you go on a plane that had a 20 percent chance of crashing?' No, of course not. It was hard to get through, though. She does put up quite a wall of words."

Meredith seemed to be working through bad news rather than drowning in it. Thoughts about unpleasant emotions aren't necessarily chatter, and this is a case in point. She didn't start spiraling. A few minutes later, after crossing another street, her verbal stream circled back to her task at hand: "Now, is there a Staples down there? I think there is."

While Meredith processed her fear about losing a beloved friend, a man named Tony fixated on another kind of grief: the loss of closeness in a relationship, and perhaps even the relationship itself. Carrying a messenger bag down a sidewalk scattered with pedestrians, he began a self-referential riff of thoughts: "Walk away . . . Look, suck it up. Or move on. Just walk away. I understand the thing about not telling everybody. But I'm not everybody. You two are having a goddamn baby. A phone call would have been good." The sense of exclusion he felt obviously cut him deeply. He seemed to be poised on a fulcrum of sorts, between a problem in search of a solution and pain that could lead to unproductive wallowing.

"Clear, totally clear. Move forward," Tony then said. He used language not just to give voice to his emotions but also to search for how best to handle the situation. "The thing is," he went on, "it could be an out. When they told me they were having a baby, I felt a bit out. I felt a bit pushed out. But now maybe it's an escape hatch. I was pissed before but, must admit, not so pissed anymore. Now it could work to my advantage." He released a soft, bitter laugh, then sighed. "I am certain that this is an out . . . I am looking at this positively now . . . I was pissed before. I felt like you two were a family . . . and you two *are* a family now. And I have an out . . . Walk tall!"

Then there was Laura.

Laura sat in a coffee shop in a restless mood. She was waiting to hear from her boyfriend, who had gone to Boston. The problem was, he was supposed to be back to help her move to a new apartment. She had been waiting for a phone call since the day before. Convinced that her boyfriend had been in a fatal accident of some sort, the night before she sat in front of her computer for four hours, every minute refreshing a keyword search of the words "bus crash." Yet, as she reminded herself, the eddy of her compulsive negative worrying wasn't just about a possible bus crash involving her boyfriend. She was in an open relationship with him, even though this wasn't something she ever desired, and it was turning out to be very hard. "It's supposed to be open for sexual freedom," she told herself, "but it's something that I never really wanted for myself . . . I don't know where he is . . . He could be anywhere. He could be with another girl."

While Meredith processed upsetting news with relative equanimity (crying at a friend's cancer diagnosis is normal) and Tony calmly coached himself to move on, Laura was stuck with repeating negative thoughts. She didn't know how to proceed. At the same time, her internal monologue dipped back in time, with reflections about the decisions that took her relationship to its current state. For her the past was very present, as was the case for Meredith and Tony. Their unique situations led them to process their experiences differently, but they were all reckoning with things that had already occurred. At the same time, their monologues also projected into the future with questions about what would happen or what they should do. This pattern of hopscotching through time and space in their inner conversations highlights something we have all noticed about our own mind: It is an avid time traveler.

While memory lane can lead us down chatter lane, there's nothing inherently harmful about returning to the past or imagining the future. The ability to engage in mental time travel is an exceedingly valuable feature of the human mind. It allows us to make sense of our experiences in ways that other animals can't, not to mention make plans and prepare for contingencies in the future. Just as we talk with friends about things we have done and things we will do or would like to do, we talk to ourselves about these same things.

Other volunteers in Irving's experiment also demonstrated preoccupations that jumped around time, braiding together in the patter of the inner voice. For example, while walking across a bridge, an older woman recalled crossing the same bridge with her father as a girl just as a man threw himself off and committed suicide. It was an indelible memory, in part because her father was a professional photographer and snapped a picture of the moment, which ended up in a citywide newspaper. Meanwhile, a man in his mid-thirties crossed the Brooklyn Bridge and thought about all the human labor it took to build it, also telling himself that he would succeed at a new job he was about to start. Another woman, waiting for a late blind date in Washington Square Park, recalled a past boyfriend who cheated on her, which ended up sparking a reverie about her desires for connection and spiritual transcendence. Other participants talked about economic hardships that might await them, while the anxieties of others centered on a looming event from a decade earlier: 9/11.

The New Yorkers who generously shared their thoughts with Andrew Irving embody the wildly diverse, richly textured nature of our default state. Their inner dialogues took them "inside" in vastly different ways, leading them down myriad streams of verbal thought. The specifics of their private conversations were as idiosyncratic as their individual lives. Yet structurally, what happened in their minds was very similar. They often dealt with negative "content," much of which sprang up through associative connections, the pinging of one thought to another. Sometimes their verbal thinking was constructive; sometimes it wasn't. They also spent a considerable amount of time thinking about *themselves,* their minds gravitating toward their own experiences, emotions, desires, and needs. The self-focused nature of the default state, after all, is one of its primary features.

The New Yorkers had these things in common, but their monologues also emphasized something else universally human: The inner voice was always there with something to say, reminding us of the inescapable need we all have to use our minds to make sense of our experiences and the role that language plays in helping us do so.

While we undoubtedly have feelings and thoughts that take nonverbal forms—visual artists and musicians, for instance, pursue precisely this kind of mental expression—humans exist in a world of words. Words are how we communicate with others most of the time (though body language and gestures are clearly instrumental too) and how we communicate with ourselves much of the time as well.

Our brain's built-in affinity for disconnecting from what is going on around us produces a conversation in our minds, one that we spend a significant portion of our waking hours engaged in. This begs a critical question: *Why?* Evolution selects qualities that provide a survival advantage. According to this rule, you wouldn't expect humans to have become such prolific self-talkers if doing so didn't add to our "fitness" for survival. But the inner voice's influence is often so subtle and fundamental that we are rarely if ever aware of all that it does for us.

The Great Multitasker

Neuroscientists often invoke the concept of neural reuse when discussing the operations of the brain—the idea that we use the same brain circuitry to achieve multiple ends, getting the absolute most from the limited neural resources at our disposal. For example, your hippocampus, the sea-horse-shaped region buried deep within your brain that creates long-term memories, also helps us navigate and move through space. The brain is a very talented multitasker. Otherwise, it would have to be the size of a bus to be large enough to support every one of its countless functions. Our inner voice, it turns out, is likewise a prodigious multitasker.

One of the brain's essential tasks is powering the engine of what is known as working memory. Humans have a natural tendency to conceptualize memory in the romantic, long-term, and nostalgic sense. We think of it as the land of the past, teeming with moments, images, and sensations that will stay with us forever and constitute our life's narrative. But then there's the fact that every minute of the day, amid an ongoing rush of stimulation that can be quite distracting (sounds, sights, smells, and so on), we have to constantly recall details to function. That we'll likely forget most of the information after it's no longer useful doesn't matter. For the brief time that information is active, we need it to function.

Working memory is what allows us to participate in work discussions and have impromptu dinner conversations. Thanks to it, we're able to remember what someone said a few seconds earlier and then incorporate it into the evolving discussion in a relevant way. Working memory is what allows us to read a menu and then order food (while also keeping up one of those conversations). It's what allows us to write an email about something urgent but not meaningful enough to get filed away in long-term storage. In short, it's what allows us to function as people out in the world. When it stops working or operates suboptimally, our capacity to perform even the most ordinary daily activities (like bugging your kids to brush their teeth while making them pack lunches and also thinking about what meetings you have later that day) fails. And connected to working memory is the inner voice.

A critical component of working memory is a neural system that specializes in managing verbal information. It's called the phonological loop, but it's easiest to understand it as the brain's clearinghouse for everything related to words that occurs around us in the present. It has two parts: an "inner ear," which allows us to retain words we've just heard for a few seconds; and an "inner voice," which allows us to repeat words in our head as we do when we're practicing a speech or memorizing a phone number or repeating a mantra. Our working memory relies on the phonological loop for keeping our linguistic neural pathways online so that we can function productively outside ourselves while also keeping our conversations going within. We develop this verbal doorway between our minds and the world in infancy, and as soon as it's in place, it propels us toward other milestones of mental development. Indeed, the phonological loop goes well beyond the realm of responding to immediate situations.

Our verbal development goes hand in hand with our emotional development. As toddlers, speaking to ourselves out loud helps us learn to control *ourselves*. In the early twentieth century, the Soviet psychologist Lev Vygotsky was one of the first people to explore the connection between language development and self-control. He was interested in the curious behavior of children who talk to themselves out loud, coaching themselves along while also doling out self-critiques. As anyone who has spent significant time around kids knows, they often have full-blown, unprompted conversations with themselves. This isn't just play or imagination; it's a sign of neural and emotional growth.

Unlike other leading thinkers of the time who thought this behavior was a sign of unsophisticated development, Vygotsky saw language playing a critical role in how we learn to control ourselves, a theory that would later be borne out by data. He believed that the way we learn to manage our emotions begins with our relationships with our primary caretakers (typically our parents). These authorities give us instructions, and we repeat those instructions to ourselves aloud, often mimicking what they say. At first, we do this audibly. Over time, though, we come to internalize their message in silent inner speech. And then later, as we develop further, we come to use our own words to control ourselves for the rest of our lives. As we all know, this doesn't mean that we always end up doing what our parents want—our verbal stream eventually develops its own unique contours that creatively direct our behavior—but these early developmental experiences influence us significantly.

Vygotsky's perspective doesn't merely explain how we learn to use our inner voice to control ourselves; it also provides us with a way of understanding how our internal conversations are "tuned" in part by our upbringing. Decades of research on socialization indicate that our environments influence how we view the world, including how we think about self-control. In families, our parents model self-control for us when we're children, and their approaches seep into our developing inner voices. Our father might repeatedly tell us never to use violence to resolve a conflict. Our mother might repeatedly tell us to never give up after a disappointment. Over time, we repeat these things to ourselves, and they begin to shape our own verbal streams.

Of course, our parents' authoritative voices are themselves shaped by broader cultural factors. For example, in most Asian countries, standing out is frowned upon, because it threatens social cohesion. In contrast, Western countries like the United States place a premium on independence, leading parents to applaud their children's individual pursuits. Religions and the values they teach likewise bleed into our household norms. In short, the voices of culture influence our parents' inner voices, which in turn influence our own, and so on through the many cultures and generations that combine to tune our minds. We are like Russian nesting dolls of mental conversations.

That said, the influence among culture, parents, and children doesn't go in just one direction. The way children behave can likewise impact their parents' voices, and we human beings of course play a role in shaping and reshaping our greater cultures as well. In a sense, then, our inner voice makes its home in us as children by going from the outside in, until we later speak from the inside out and affect those around us.

Recent research that Vygotsky didn't live to see has taken his theory further, with studies demonstrating that children brought up in families with rich communication patterns develop this facet of inner speech earlier. Moreover, it turns out that having imaginary friends may spur internal speech in children. In fact, emerging research suggests that imaginary play promotes self-control, among many other desirable qualities such as creative thinking, confidence, and good communication.

Another crucial way the inner voice helps us control ourselves is by evaluating us as we strive toward goals. Almost like a tracking app on a phone, the default state monitors us to see if we're meeting benchmarks at work to get that end-of-the-year raise, if we're advancing on our side-hustle dream of opening a restaurant, or if our relationship with that friend we have a crush on is developing apace. This frequently happens with a verbal thought popping in our mind much like an appointment reminder appearing on your lock screen. In fact, spontaneous thoughts related to goals are among the most frequent kind that fill our mind. It's our inner voice alerting us to pay attention to an objective.

Part of reaching goals involves making the right choice

when there is a proverbial fork in the road, which is why our inner voice also allows us to run mental simulations. For example, when we are engaged in creative brainstorming about, say, the best way to do a presentation or the best melodic progression for a song we're writing, we internally explore different possible paths. Often even before writing the words for a presentation or touching a musical instrument, we've already tapped our introspective capacities to decide on the best permutation. The same goes for figuring out how to deal with an interpersonal challenge, the way Tony did while walking around New York thinking about his friends who hadn't told him about their pregnancy. He was simulating whether he should remain close or distance himself. This multiple-reality brainstorming even happens while we are sleeping, in our dreams.

Historically, psychologists thought of dreams as a chamber of their own in the mind, and very different from what happens during our waking hours. Freud, of course, thought dreams were the royal road to the unconscious, a locked box holding our repressed urges, and psychoanalysis was the key that opened it. With our defenses down and our civilized propriety turned off while we slept, he thought, our demons came out and romped around, revealing our desires. Then came early neuroscience, which took out all the dark and naughty romance of psychoanalysis and replaced it with the cold no-nonsense attitude of the physical workings of the brain. It said that dreams were nothing more than the brain's way of interpreting random brain-stem firings during REM sleep. Out the door went sexual symbolism, which was entertaining if a bit loony, and in came the mechanics of neurons, which was more scientifically grounded (and not at all salacious).

Present-day research with more advanced technology has

shown that our dreams in fact share many similarities with the spontaneous verbal thoughts we experience when we are awake. It turns out that our waking verbal mind converses with our sleeping one. Fortunately, this doesn't produce Oedipal wish fulfillments.

It can help us.

Emerging evidence suggests that dreams are often functional and highly attuned to our practical needs. You can think of them as a slightly zany flight simulator. They aid us in preparing for the future by simulating events that are still to come, pointing our attention to potentially real scenarios and even threats to be wary of. Although we still have much to learn about how dreams affect us, at the end of the day—or night, rather—they are simply *stories* in the mind. And sure enough, in waking life, the inner voice pipes up loudly about the most foundational psychological story of all: our identities.

Our verbal stream plays an indispensable role in the creation of our selves. The brain constructs meaningful narratives through autobiographical reasoning. In other words, we use our minds to write the story of our lives, with us as the main character. Doing so helps us mature, figure out our values and desires, and weather change and adversity by keeping us rooted in a continuous identity. Language is integral to this process because it smooths the jagged and seemingly unconnected fragments of daily life into a cohesive through line. It helps us "storify" life. The words of the mind sculpt the past, and thus set up a narrative for us to follow into the future. By flitting back and forth between different memories, our internal monologues weave a neural narrative of recollections. It sews the past into the seams of our brain's construction of our identity.

The brain's multitasking abilities are varied and vital, as is the inner voice. But to truly understand its profound value, we must contemplate what it would be like if our verbal thoughts were to disappear. As improbable as this may sound, we don't have to merely imagine this scenario. In some cases, it actually happens.

Going to La-La Land

On December 10, 1996, Jill Bolte Taylor woke just as she did every morning. A thirty-seven-year-old neuroanatomist, she worked in a psychiatry lab at Harvard University, where she studied the makeup of the brain. Her drive to map our cortical landscapes to understand their cellular interactions and the behaviors they produced grew out of her family history. Her brother had schizophrenia, and though she couldn't expect to reverse his illness, it motivated her to try to unravel the mysteries of the mind. She was well on her way to doing so—that is, until the day her brain stopped functioning well.

Bolte Taylor got out of bed to do her morning exercise on a cardio machine, but she didn't feel like herself. She had a pulsing pain behind her eye, like an ice-cream headache that came and went, and came and went. Then, once she started exercising, things got strange. While on the machine, she felt her body slow down and her perception contract. "I can no longer define the boundaries of my body," she later recalled. "I can't define where I begin and where I end."

Not only did she lose the sense of her body in physical space, she also began to lose her sense of who she was. She felt her emotions and memories drift away, as if they were leaving her to take up residence elsewhere. The second-by-second sparking of perceptions and reactions that characterized her normal mental awareness faded. She felt her thoughts losing their shape, and with them, *her words*. Her verbal stream slowed like a river drying up. Her brain's linguistic machinery broke down.

A blood vessel had popped on the left side of her brain. She was having a stroke.

While her physical movements and linguistic faculties were drastically encumbered, she managed to phone a colleague, who quickly gathered that something was wrong. Soon after, Bolte Taylor found herself in the back of an ambulance being taken to Massachusetts General Hospital. "I felt my spirit surrender," she said. "I was no longer the choreographer of my life." Sure that she was going to die, she said farewell to her life.

She didn't die. Later that afternoon she woke up in a hospital bed, astonished that she was still alive, though her life wouldn't be the same for a long time. Her inner voice as she had always known it had departed. "My verbal thoughts were now inconsistent, fragmented, and interrupted by an intermittent silence," she later recalled. "I was alone. In the moment, I was alone with nothing but the rhythmic pulse of my beating heart." She wasn't even alone with her thoughts, because she didn't *have* thoughts as she'd had before.

Her working memory wasn't working, making it impossible to complete the simplest tasks. Her phonological loop, it seemed, had unraveled. Her self-talk was silenced. She was no longer a mental time traveler capable of revisiting the past and imagining the future. She felt vulnerable in a way she had never even imagined possible, as if she were spinning by herself in outer space. She wondered, wordlessly, if words would ever return in full to her mental life. Without verbal introspection, she ceased to be human in the previous sense she had known. "Devoid of language and linear processing," she wrote, "I felt disconnected from the life I had lived."

Most profoundly of all, she lost her identity. The narrative her inner voice had allowed her to construct over nearly four decades erased itself. "Those little voices inside your head," as she put it, had made her *her,* but now they were silent. "So, was I really still me? How could I still be Dr. Jill Bolte Taylor, when I no longer shared her life experiences, thoughts, and emotional attachments?"

When I imagine what it would be like to go through what Jill Bolte Taylor experienced, it fills me with panic. Losing the ability to talk to myself, to use language to tap into my intuitions, stitch together my experiences into a coherent whole, or plan for the future, sounds much worse than a letter from a deranged stalker. Yet it's here where her story gets stranger and even more fascinating.

Bolte Taylor wasn't frightened the way I imagine I or anyone else in her situation would feel. Remarkably, she found a comfort like nothing she had ever felt before when her lifelong internal conversation vanished. "The growing void in my traumatized brain was entirely seductive," she later wrote. "I welcomed the reprieve that the silence brought from the constant chatter."

She had gone, as she put it, to "la-la land."

Being robbed of language and memory, on the one hand, was terrifying and lonely. On the other hand, it was ecstatically, euphorically liberating. Free from her past identity, she could also be free from all her recurring painful recollections, present stresses, and looming anxieties. Without her inner voice she was free from chatter. To her, this trade-off felt worth it. She later reflected that this was because she hadn't learned to manage her buzzing inner world prior to her stroke. Like all of us, she had trouble controlling her emotions when she got sucked into negative spirals.

Two and a half weeks after her stroke, Bolte Taylor would have surgery to remove a golf-ball-sized blood clot from her brain. It would take her eight years to fully recover. She continues to conduct research on the brain while also sharing her story with the world. She emphasizes the overwhelming sense of generosity and well-being she felt when her inner critic was muted. She is now, as she describes it, "a devout believer that paying attention to our self-talk is vitally important for our mental health."

What her experience shows us in singularly vivid terms is how deeply we struggle with our inner voice—to the point where the stream of verbal thoughts that allows us to function and think and be ourselves could lead to expansively good feelings when it's gone. This is striking evidence of how influential our inner voice can be. Research bears out this phenomenon in less exceptional circumstances. Not only can our thoughts taint experience. They can blot out nearly everything else.

A study published in 2010 drives home this point. The scientists found that inner experiences consistently dwarf outer ones. What participants were thinking about turned out to be a better predictor of their happiness than what they were actually doing. This speaks to a sour experience many people have had: You're in a situation in which you should be happy (spending time with friends, say, or celebrating an accomplishment), but a ruminative thought swallows your mind. Your mood is defined not by what you did but by what you thought about.

The reason people experience relief when their inner voice quiets isn't that it is a curse of our evolution. As we've seen, we have a voice in our heads because it is a unique gift that accompanies us from the streets of New York to our sleeping dreams. It allows us to function in the world, achieve goals, create, connect, and define who we are in wonderful ways. But when it morphs into chatter, it is often so overwhelming that it can cause us to lose sight of this and perhaps even wish we didn't have an inner voice at all.

Before we get into what science teaches us about how to control our verbal mental stream, though, we need to understand the harmful effects of chatter that require us to intervene in the first place. When you take a close look at what our destructive verbal thoughts can do to us—to our minds, bodies, and relationships—you realize that shedding a few tears on the streets of New York is getting off easy.

CHAPTER TWO

When Talking to Ourselves Backfires

The first wild pitch seemed like a fluke.

It was October 3, 2000, game one between the St. Louis Cardinals and the Atlanta Braves in the first round of the National League playoffs. The Cardinal pitcher Rick Ankiel watched the ball he had just thrown bounce off the ground past his catcher and then hit the backstop. As the runner on first jogged to second, the crowd made a sound of modest, almost supportive surprise—he was, after all, playing on his home turf at Busch Stadium in St. Louis—though there was no reason to think his wild throw presaged any shift in the balance of the inning. In baseball, pitches occasionally get away from even the best pitchers, never mind the fact that Ankiel wasn't just any pitcher.

When he was drafted right out of high school, a seventeen-year-old with a ninety-four-mile-an-hour fastball, scouts and commentators believed Ankiel had the potential to be one of the best pitchers the game had seen in decades. His debut in the majors two years later didn't disappoint. During his first full season in 2000, he struck out 194 batters, racking up eleven wins to help his team reach the playoffs. Everything pointed to a spectacular career. It was no surprise, then, that he was chosen as the starter for game one against the Braves in the playoffs that October. All he had to do was the thing he did best in life: throw a baseball.

Ankiel tried to forget the wild pitch. It was an anomaly for him, and there was nothing to worry about. It was only the third inning and his team had already jumped to a dramatic 6–0 lead. On top of that, the pitch hadn't even been that wild; it had just ricocheted off the ground the wrong way and gotten away from his catcher. He'd felt good going into the inning, so he would just shake it off. And yet a prickly nettle of a thought lodged itself in his mind as he gathered himself on the mound. *Man,* he said to himself, *I just threw a wild pitch on national TV.* What he didn't know was that he *did* have something to worry about.

Moments later, after reading the signs from his catcher, Ankiel uncoiled his explosive, left-handed windup and . . . threw another wild pitch.

The crowd oohed a bit louder and longer this time, as if sensing something were off. The runner on second ran to third base. While the dark-eyed, twenty-one-year-old Ankiel chewed his gum and kept an unreadable facial expression, inside he was anything but composed. As his catcher retrieved the ball again and the seconds passed beneath the afternoon sun, he felt his mind slipping out of his control and into the hands of what he would come to call "the monster"—his cruel inner critic, a stream of verbal thoughts so vicious they could undo years of hard work, its voice louder than the fifty-two thousand fans in the stands.

Anxiety. Panic. Fear.

His own immense vulnerability—a young player with everything on the line—was something he could no longer ignore.

Ankiel might have looked like the shining embodiment of the American dream—a small-town kid from Florida making good on his exceptional gift—but his childhood belied such a picturesque narrative. The son of a verbally and physically abusive father who was both a petty criminal and an addict, he knew depths of emotional pain beyond his years. Which is why baseball was more than just a career for him. It was a hallowed, safe place where he felt good, where things were easy, where a kind of joy was built in, unlike his family life. Only now something strange and seemingly uncontrollable was starting to happen, overwhelming his senses and flooding him with terror.

Still, he was determined to rally. He narrowed his focus in on his weight, on his stance, on his arm. All he had to do was make the machinery of his windup click into place. Then he wound up again.

And threw another wild pitch.

And another.

And another.

Before the Cardinals gave up any more runs, Ankiel was pulled out of the game. He disappeared into the dugout accompanied by "the monster" inside him.

His showing on the mound that day was both embarrassing and unexpected. The last time a pitcher had thrown five wild pitches in one inning had been more than a hundred years earlier. But it wouldn't have retrospectively gone down as one of the most painful-to-watch performances in baseball history were it not for what soon followed.

When Ankiel was called on to pitch against the Mets nine

days later, the same thing happened. The monster reappeared and he threw more wild pitches. Once again, he was pulled from the mound, this time before the first inning was over. And yet the humiliation didn't end there, though his brief career as a major-league pitcher effectively did.

At the start of the following season, Ankiel pitched a few more games during which he had to drink alcohol to stay his nerves before taking the field, but even the liquor couldn't help calm his mind. His pitching didn't improve. He was sent to the minors, where he spent a dispiriting three years before deciding to retire from baseball in 2005 at the tragically premature age of twenty-five.

"I can't do this anymore," he told his coach.

Rick Ankiel would never pitch professionally again.

Unlinking and the Magical Number Four

Rick Ankiel isn't the first elite athlete to lose his superpower—to suddenly have the skill he was best at stop being a skill altogether. Time and again, people who have spent years mastering a talent watch it break down like a decrepit old Chevy when chatter hijacks their inner voice. This phenomenon isn't restricted to athletes. It can happen to anyone who has become skilled at a learned task—from teachers who memorize their lesson plans, to start-up founders with rehearsed spiels they pitch to investors, to surgeons who perform complex operations that took them years to master. The explanation for why these skills fail ultimately relates to how the conversations we have with ourselves influence our *attention*.

At any given moment, we are bombarded with information—countless sights and sounds, and the thoughts and feelings that these stimuli spark. Attention is what allows us to filter out the things that don't matter so that we can focus on the things that do. And although much of our attention is involuntary, like when we automatically turn toward a loud noise, one of the features that make humans so unique is our ability to consciously concentrate on the tasks that require our attention.

When we find ourselves overwhelmed by emotion, as Ankiel did on that fall day in 2000, one of the things our inner voice does is harness our attention, narrowing it in on the obstacles we encounter to the exclusion of practically everything else. This serves us well most of the time, but *not* when it comes to exercising our attention to wrangle an automatic, learned skill, as pitching was for Ankiel. To understand why this is, it's useful to look at what goes right when athletes' automatized behaviors lift them into the most impressive heights of performance.

On August 11, 2019, the American gymnast Simone Biles made sports history when she became the first woman to ever land a triple-double flip at an official competition during her floor routine at the U.S. Gymnastics Championships. As one commentator wrote, "It's a move that requires incredible, almost superhuman strength, coordination, and training." To execute it by deliberately thinking about each movement would be impossible, because everything happens in the air, where the laws of physics play out in an instant—gravity versus a body.

The seemingly impossible move Biles pulled off required her to spin her body around two axes at the same time and do two backflips while also spinning three times, hence the name triple double. We can think of her perfect execution of the move as the culmination of all the automatized movements her brain had mastered over the years: running, jumping, hand-springs, backflips, twists, and landing. To achieve her triple double, she linked into one breathtaking feat a set of movements that took years to learn but that eventually stopped needing her brain's conscious control. Biles's inner voice didn't direct her every action, though it likely rejoiced as the crowd went wild.

Like all athletes, Biles built her triple double out of a series of individual behaviors that she linked together through practice. Eventually, the separate elements in the chain of movements merged into one seamless action. Her automatic bodily mechanics, spurred by her brain's ability to link them together (combined with fabulous DNA), propelled Biles into sports history. Until Ankiel's meltdown, he seemed as if he were on a similar trajectory, with flawless movements and a preternaturally strong arm. So, what happened that day on the mound?

He *un*linked.

Ankiel's verbal stream turned into a spotlight that shined his attention too brightly on the individual physical components of his pitching motion, thereby seeming to inadvertently dismantle it. After throwing the first few wild pitches, he mentally stepped back and focused on the mechanics of his throw: the choreographed movements that involved his hips, legs, and arm. On the surface, that seems wise and intuitive. He was calling on his brain to fix a scripted behavior it had previously successfully carried out literally tens of thousands of times. Which is precisely where things went wrong.

When you're completing your taxes, it pays to double-check your calculations to make sure you've done everything right, even if you're an experienced accountant. But for well-worn, automatic behaviors that you're trying to execute under pressure, like pitching, this very same tendency leads us to break down the complicated scripts that we've learned to execute without thinking. This is exactly what our inner voice's tendency to immerse us in a problem does. It overfocuses our attention on the parts of a behavior that only functions as the *sum* of its parts. The result: paralysis by analysis.

Chatter ruined Ankiel's career as a pitcher, but automatic behaviors aren't the only kind of skill that can backfire when our inner voice betrays us. After all, one of the things that distinguishes us from every other animal species is our ability not just to execute automatic behaviors but to use our mind to consciously focus our attention.

It is our ability to reason logically, solve problems, multitask, and control ourselves that allows us to manage work, family, and so many other crucial parts of our lives with wisdom, creativity, and intelligence. To do this, we have to be deliberate and attentive and flexible, which we are capable of doing thanks to what we can think of as the CEO of the human brain—our *executive functions,* which are also vulnerable to the incursions of an unsupportive inner voice.

Our executive functions are the foundation of our ability to steer our thoughts and behavior in the ways we desire. Supported largely by a network of prefrontal brain regions located behind our forehead and temples, their job is to intervene when our instinctual processes aren't sufficient and we need to consciously guide our behavior. They allow us to keep relevant information active in our mind (working memory is a part of executive functions), filter out extraneous information, block out distractions, play with ideas, point our attention where it needs to go, and exercise self-control—like helping us resist the temptation to open a new browser tab and go down a tangential Wikipedia rabbit hole. In short, without our executive functions we wouldn't be able to function in the world.

The reason your brain needs this type of neurological leadership is that paying attention, reasoning wisely, thinking creatively, and executing tasks often require you to leave automatic mode and exercise conscious effort. And doing this asks a lot of your executive functions because they have a limited capacity. Like a computer that slows down when it has too many programs open, your executive functions perform worse as the demands placed on them increase.

The classic illustration of this limited capacity, known as the magical number four, has to do with our ability to hold between three and five units of information in the mind at any given time. Take an American phone number. Memorizing the number 200-350-2765 is much easier than memorizing 2003502765. In the first instance, you've grouped the numbers, so you're memorizing three pieces of information; in the latter, you're trying to memorize an unbroken string of ten pieces of information, placing more demands on the system.

Your labor-intense executive functions need every neuron they can get, but a negative inner voice hogs our neural capacity. Verbal rumination concentrates our attention narrowly on the source of our emotional distress, thus stealing neurons that could better serve us. In effect, we jam our executive functions up by attending to a "dual task"—the task of doing whatever it is we want to do *and* the task of listening to our pained inner voice. Neurologically, that's how chatter divides and blurs our attention.

All of us are familiar with the distractions of a negative verbal stream. Have you ever tried to read a book or complete a task requiring focus after a bad fight with someone you love? It's next to impossible. All the resulting negative thoughts consume your executive functions because your inner critic and its ranting have taken over corporate headquarters, raiding your neuronal resources. The problem for most of us, however, is that usually we're engaged in activities with much higher stakes than retaining information in a book. We're doing our jobs, pursuing our dreams, interacting with others, and being evaluated.

Chatter in the form of repetitive anxious thought is a marvelous saboteur when it comes to focused tasks. Countless studies reveal its debilitating effects. It leads students to perform worse on tests, produces stage fright and a tendency to catastrophize among artistic performers, and undermines negotiations in business. One study found, for instance, that anxiety led people to make low initial offers, exit discussions early, and earn less money. This is a very nice way of saying they failed at their jobs—because of chatter.

On any given day, the keel of our inner voice can be thrown askew by an infinite number of things. When this happens, we have trouble focusing our minds to address the inevitable daily challenges we face, which often produces still more turbulence in our inner dialogues. Quite naturally, when we're floundering like this, we look for a way out of our predicament. So, what exactly do we do?

That's the question a middle-aged, mild-mannered psychologist became intrigued with some thirty years ago. His research would raise profound questions about the costs of chatter that go far beyond our ability to focus our attention. Our inner voice affects our social lives as well.

A Social Repellent

In the late 1980s, a bespectacled Belgian psychologist named Bernard Rimé decided to examine whether experiencing the kinds of strong negative emotions that characterize chatter lead people to engage in a very social process: talking.

Over the course of several studies, Rimé brought people into his laboratory and asked them whether they talked about negative experiences from their past with others. Then he turned his focus to the present and asked people to record in diaries over the course of several weeks each time they confronted an upsetting situation and whether they discussed it with members of their social networks. He also ran experiments in which he provoked participants in the lab and then watched if they shared their reactions with others nearby.

Again and again, Rimé landed on the same finding: People feel compelled to talk to others about their negative experiences. But that wasn't all. The more intense the emotion was, the more they wanted to talk about it. Additionally, they returned to talking about what had occurred more often, doing so repeatedly over the course of hours, days, weeks, and months, and sometimes even for the remainder of their lives.

Rimé's finding proved true regardless of people's age or education level. It was characteristic of men as well as women. It even carried across geography and cultures. From Asia to the Americas to Europe, he kept finding the same thing: Strong emotions acted like a jet propellant, blasting people off to share their experiences. It seemed to be a law of human nature. The only exceptions to this rule were cases in which people felt shame, which they often wished to conceal, or certain forms of trauma, which they wanted to avoid dwelling on.

Such consistency in a finding was stunning, though it may

sound like a confirmation of the obvious. As we all know, people talk a lot about intense emotions. It's not as if we go around calling friends to say, "Hey, I feel fairly normal today." It's the highs and lows that leap from the verbal stream in our minds to the words that leave our mouths.

While this sounds normal and harmless, *repeatedly* sharing our negative inner voice with others produces one of the great ironies of chatter and social life: We voice the thoughts in our minds to the sympathetic listeners we know in search of their support, but doing so excessively ends up pushing away the people we need most. It's as though the pain of chatter makes people less sensitive to the normal social cues that tell us when enough is enough. To be clear, this doesn't mean that talking to others about your problems is harmful per se. But it highlights how chatter can transform an otherwise helpful experience into something negative.

Many of us have a limited threshold for how much venting we can listen to, even from the people we love, as well as how often we can tolerate this venting while not feeling listened to ourselves. Relationships thrive on reciprocity. That's one of the reasons why therapists charge us for their time and friends don't. When this conversational balance becomes lopsided, social connections fray.

To make matters worse, when this occurs, the people who are overventing and inadvertently alienating those around them are less capable of solving problems. This makes it harder for them to fix the breach in their relationships, begetting a vicious cycle that can end with a toxic outcome: loneliness and isolation.

For a heightened example of how this process of progressive social isolation operates, we can look to that widespread emotional tumult known as middle school. One study tracked more than one thousand middle schoolers for seven months and found that kids who were prone to rumination reported talking with their peers more than their low-rumination counterparts. Yet this did more harm than good. It predicted a host of painful results: being socially excluded and rejected, being the target of gossip and rumors by their peers, and even being threatened with violence.

Unfortunately, in this case, what is true of preteens and teens crosses over to adulthood. Furthermore, it turns out it doesn't matter much even if you have a legitimate reason for venting; overvoicing your chatter can still push people away. One study that focused on grieving adults found that people who were prone to ruminate reached out for more social support after their loss, which is normal. The uncomfortable twist, though, is that they reported experiencing more social friction and less emotional support in their relationships as a result.

Uncontrolled emotional sharing isn't the only type of social repellent that chatter enables. People who perseverate on conflicts are also more likely to behave aggressively. One experiment showed that cuing people to ruminate about how they felt after being insulted by an experimenter who undiplomatically criticized an essay they had written led them to be more hostile with the person who insulted them. When given the chance to administer loud blasts of noise to the experimenter, they did so more than people who didn't ruminate. In other words, the more I stew over what you did to me, the more I keep those negative feelings alive, and the more likely I am to act aggressively against you as a result. Chatter also leads us to displace our aggression against people when they don't deserve it. Our boss upsets us, for example, and we take it out on our kids.

But none of this research considers our *digital* lives. In the age of online sharing, Rimé's work on emotions and our social lives has acquired a new urgency. Facebook and other social media applications like it have provided us with a world-altering platform for sharing our inner voice and listening in on the inner voices of others (or at least what other people want us to think they're thinking about). Indeed, the first thing people see when they log in to Facebook is the prompt asking them to broadcast their answer to this question: *"What's on your mind?"*

And broadcast we do.

In 2020, close to two and a half billion people use Facebook and Twitter—almost a third of the world's population—and they frequently do so to share their private ruminations. It's worth highlighting that there's nothing inherently bad about sharing on social media. In the long historical timeline of our species, it's simply a new environment that we find ourselves spending a great deal of time in, and environments aren't good or bad per se. Whether they help or harm us depends on how we interact with them. That said, there are two features of social media that are worrying when you consider the intense drive we have to air our stream of thoughts: empathy and time.

It's hard to overstate the importance of empathy both individually and collectively. It's what allows us to forge meaningful connections with others, it's one of the reasons why we so often find ourselves venting (we seek the empathy of others), and it's also one of the mechanisms that holds communities together. It's a capacity we evolved because it helps our species survive.

Research shows that observing other people's emotional responses—seeing someone wince or hearing a quiver in a voice—can be a potent route to triggering empathy. But on-line, the subtle physical gestures, micro-expressions, and vocal intonations that elicit empathic responses in daily life are ab-sent. As a result, our brains are deprived of information that serves a critical social function: inhibiting cruelty and antiso-cial behavior. In other words, less empathy all too frequently leads to trolling and cyberbullying, which have grave conse-quences. Cyberbullying, for example, has been linked to lon-ger episodes of depression, anxiety, and substance abuse, as well as several toxic physical effects like headaches, sleep distur-bances, gastrointestinal ills, and changes in how well stress-response systems operate.

The passage of time is likewise essential to helping us man-age our emotional lives, especially when it comes to processing upsetting experiences. When we identify someone to talk to off-line, we often have to wait until we see the person or until they're available to chat. While one waits for that person, something magical happens: Time passes, which allows us to reflect on what we're feeling and thinking about in ways that often temper our emotions. Indeed, research supports the com-mon idea that "time heals" or the advice to "just give it time."

Now let's transplant ourselves to the parallel world of digi-tal life and our ability to access it anytime thanks to our smart devices. Social media allows us to connect with others in the immediate aftermath of a negative emotional response, before time provides us with the opportunity to rethink how we're feeling or what we're planning to do. Thanks to twenty-first-century connectivity, during the very peak of our inner flare-ups, right when our inner voice wants to rant from the rooftops, it can.

We post. We tweet. We comment.

With the passage of time and physical elicitors of empathy removed, social media becomes a place amenable to the unseemly sides of the inner voice. This can lead to increased conflict, hostility, and chatter for both individuals and arguably society as a whole. It also means that we overshare more than ever before.

Similar to talking for too long and too frequently to others about your problems, overly emotional posts irritate and alienate others. They violate unspoken norms, and users wish people who overshare online would look for support from friends off-line. Unsurprisingly, people with depression—which is fueled by the verbal stream—share more negative personal content on social media yet actually perceive their network as less helpful than nondepressed people do.

But social media doesn't just provide us with a platform to (over)share the thoughts and feelings streaming through our head, and the ways it derails our internal dialogues don't exclusively relate to empathy and time. Social media also allows us to shape what we want other people to believe is happening in our lives, and our choices about what to post can fuel other people's chatter.

The human need to self-present is powerful. We craft our appearances to influence how people perceive us all the time. This has always been the case, but then along came social media to give us exponentially more control over how we do this. It allows us to skillfully curate the presentations of our lives—the proverbial photoshopped version of life, with the low points and less aesthetically pleasing moments left out. Engaging in this self-presentation exercise can make us feel better, satisfying our own need to appear positively in the eyes of others and buoying our inner voice.

But there's a catch. Although posting glamour shots of our lives may lead us to feel better, that very same act can cause the users who view our posts to feel worse. That's because at the same time that we are motivated to present ourselves positively, we are also driven to compare ourselves with others. And social media switches the social-comparison hardware in our brain into overdrive. A study my colleagues and I published in 2015 demonstrated, for example, that the more time people spent passively scrolling through Facebook, peering in on the lives of others, the more envy they experienced and the worse they subsequently felt.

If broadcasting our feelings on social media and participating in its culture of self-curation have so many chatter-inducing effects, it's reasonable to ask why we continue to share. One answer to that question has to do with the trade-off that often comes from engaging in behaviors that feel good in the moment but have negative consequences over time. Research shows that the same brain circuitry that becomes active when we are attracted to someone or consume desirable substances (everything from cocaine to chocolate) also activates when we share information about ourselves with others. In a particularly compelling illustration, one study by Harvard neuroscientists published in 2012 showed that people would prefer to share information about themselves with others than receive money. The social high, in other words, is like a neuronal high, a delicious hit for our dopamine receptors.

The point of all this is to say that both online and off-line, when we let our chatter drive social behaviors, we frequently crash into a range of negative outcomes. The most damaging one for internal and external conversations is that, quite often, we end up finding less support. This starts a vicious cycle of social isolation, which further wounds us. In fact, if you stop and listen, you'll notice that many people actually use the language of physical "pain" to describe how they feel when they're rejected by others.

In languages across the globe, from Inuktitut to German, Hebrew to Hungarian, Cantonese to Bhutanese, people use words related to physical injury to describe emotional pain— "damaged," "wound," "injured," among many others. It turns out the reason they do so is not just that they have a knack for metaphorical expression. One of the most chilling discoveries I've had in my career is that chatter doesn't simply hurt people in an emotional sense; it has physical implications for our body as well, from the way we experience physical pain all the way down to the way our genes operate in our cells.

The Piano Inside Our Cells

One by one they arrived in our basement laboratory: the heartbroken of New York City.

It was 2007. My colleagues and I had begun a study to better understand what emotional pain really looked like in the brain. Instead of finding just any volunteers to participate— which would have meant finding a method for making them feel bad in the lab in some effective but somehow still ethical way—we sought out forty volunteers who were already hurting: people who had recently suffered heartbreak, one of the most potent elicitors of emotional torment that we know. We posted ads in the subway and in parks looking for people who had just been rejected from monogamous relationships that had lasted at least six months:

HAVE YOU RECENTLY HAD A DIFFICULT, UNWANTED BREAKUP?

STILL HAVE FEELINGS FOR AN EX-PARTNER?

PARTICIPATE IN AN EXPERIMENT ON HOW THE BRAIN
PROCESSES EMOTIONAL AND PHYSICAL PAIN!

In a city of eight million, volunteers were easy to find.

We did do one slightly provocative thing, though. We had them bring a photo of the person who'd dumped them. Having the photos wasn't gratuitous. By asking the volunteers to lie in an MRI scanner, look at the object of their unrequited love, and recall how they felt during the precise moment of their breakup, we were hoping to obtain a neural snapshot of chatter. But we also wanted to know something else: whether the way the brain processes an experience of emotional pain was similar to how it processes *physical* pain. To get at the latter, we also applied heat to their arms that felt like a hot cup of coffee.

Afterward, we compared the MRI results from when they looked at the photo of their lost love with those of the hot-coffee simulation. Incredibly, there was a high degree of overlap in brain regions that play a role in our sensory experience of physical pain. In other words, our results suggested that emotional pain had a physical component as well.

These and a host of findings from other labs that emerged around the same time were beginning to demonstrate how admittedly fuzzy concepts like social pain influence what happens in our bodies, especially when it comes to stress.

It's a cliché of the twenty-first century to say that stress kills. It's a modern epidemic that contributes to productivity losses in the United States alone that amount to $500 billion annually. Yet we frequently lose sight of the fact that stress is an adaptive response. It helps our bodies respond quickly and efficiently to potentially threatening situations. But stress stops being adaptive when it becomes *chronic*—when the fight-or-flight alarm fails to *stop* signaling. And sure enough, a main culprit in keeping stress active is our negative verbal stream.

Threat includes physical danger, of course, but it also encompasses a range of more common experiences. For example, when we encounter situations that we aren't sure we can handle: losing a job or starting a new one, having a conflict with a friend or family member, moving to a new city, facing a health challenge, grieving the death of a loved one, getting a divorce, living in an unsafe neighborhood. These are all adverse circumstances capable of triggering a threat response similar to the one we get when we are in immediate physical danger. When the threat trip wire in our brain gets crossed, our bodies quickly mobilize themselves to protect us, much like a country mobilizing its army for a coordinated strike against an enemy invader.

Phase one begins instantly in a cone-shaped region of the brain called the hypothalamus. When your hypothalamus receives signals from other parts of your brain indicating that threat exists, it triggers a chain of chemical reactions that release adrenaline into your bloodstream. The adrenaline causes your heart to beat fast, your blood pressure and energy levels to rise, and your senses to sharpen. Moments later, the stress hormone cortisol is released to keep your jet engines firing and maintain your energy levels. While all of this is happening, chemical messengers are also working to curb the systems in your body that aren't vital to your ability to respond to an immediate threat, like your digestive and reproductive systems. If you've ever noticed your appetite for food or sex disappear when you're in the midst of a crisis, these chemical messengers are the reason why. All of these changes have a singular goal: to enhance your ability to respond quickly to the stressors you face, regardless of whether you're actively confronting those stressors in the moment (like seeing a burglar enter your home) or simply conjuring them up in your mind.

Yes, we can create a chronic physiological stress reaction just by thinking. And when our inner voice fuels that stress, it can be devastating to our health.

Countless studies have linked the long-term activation of our stress-response systems with illnesses that span the gamut from cardiovascular disease to sleep disorders to various forms of cancer. This explains how stressful experiences such as feeling chronically isolated and alone can have drastic effects on our health. Indeed, not having a strong social-support network is a risk factor for death as large as smoking more than fifteen cigarettes a day, and a greater risk factor than consuming excessive amounts of alcohol, not exercising, being obese, or living in a highly polluted city.

Chronic negative thoughts can also push into the territory of mental illness, though this isn't to say chatter is the same thing as clinical depression, anxiety, or post-traumatic stress disorder. Repetitive negative thinking isn't synonymous with these conditions, but it's a common feature of them. Indeed, scientists consider it a *transdiagnostic risk factor* for many disorders, meaning that chatter underlies a variety of mental illnesses.

But here is what is most frightening about the ways in which chatter feeds stress. When our panic response is prolonged, the gradual physiological erosion it causes can harm more than our ability to fight sickness and keep our body running smoothly. It can change the way our DNA influences our health.

When I was in college, I learned a simple formula: Genes + Environment = Who We Are. In class after class, my professors told me that when it came to the shaping of human life, the effects of genes and environment didn't mix. Nurture was in one box, and Nature was in another. This was conventional wisdom for a long time—until suddenly it wasn't. To many a scientist's surprise, new research suggests that this equation couldn't be further from the truth. Just because you have a certain type of gene doesn't mean it actually affects you. What determines who we are is whether those genes are turned on or off.

One way to think about this is to imagine that your DNA is like a piano buried deep in your cells. The keys on the piano are your genes, which can be played in a variety of ways. Some keys will never be pressed. Others will be struck frequently and in steady combinations. Part of what distinguishes me from you and you from everyone else in the world is how these keys are pressed. That's gene expression. It's the genetic recital within your cells that plays a role in forming how your body and mind work.

Our inner voice, it turns out, likes to tickle our genetic ivories. The way we talk to ourselves can influence which keys get played. The UCLA professor of medicine Steve Cole has spent his career studying how nature and nurture collide in our cells. Over the course of numerous studies he and his colleagues discovered that experiencing chatter-fueled chronic threat influences how our genes are expressed.

Cole and others have found that a similar set of inflammation genes are expressed more strongly among people who experience chronic threat, regardless of whether those feelings emerge from feeling lonely or dealing with the stress of poverty or the diagnoses of disease. This happens because our cells interpret the experience of chronic *psychological threat* as a viscerally hostile situation akin to being physically attacked.

When our internal conversations activate our threat system frequently over time, they send messages to our cells that trigger the expression of inflammation genes, which are meant to protect us in the short term but cause harm in the long term. At the same time, the cells carrying out normal daily functions, like warding off viral pathogens, are suppressed, opening the way for illnesses and infections. Cole calls this effect of chatter "death at the molecular level."

Asset or Liability?

Learning about the effects that our negative internal dialogues can have on our minds, relationships, and bodies can be deeply unsettling. As a scientist steeped in this work, I often can't help but think how this research applies to my own life and the lives of those I love. I'd be lying if I told you I didn't worry each time I see one of my daughters fretting over something.

And yet, if I look around me, I see examples that offer hope. I see students who go from insecure freshmen drowning in self-doubt to confident seniors ready to make contributions to the world. I see people who face tremendous hardships find ways to connect with others and receive support from their social networks. And I see those who have lived with chronic stress attain healthy lives. As a young woman in Poland, my grandmother Dora escaped the Nazis by hiding in the forest a whole terrifying year, and yet she still managed to live seventy more resilient, joyful years in the United States.

What these important counterexamples bring me back to is that great puzzle of the human mind: how our inner voice can be both a liability and an asset. The words streaming through our heads can unravel us, but they can also drive us toward meaningful accomplishments . . . if we know how to control them. At the same time that our species evolved the inner voice, which can drown us in chatter, we also co-evolved tools to turn it into our greatest strength. Just look at Rick Ankiel, who returned to the major leagues in 2007—not as a pitcher, but as an outfielder who still had to contend with the pressures of playing in front of tens of thousands of fans.

Ankiel would play in the majors for another seven years, known for his rocket arm in the outfield and his explosive swing at the plate. He was the pitcher who'd lost his career, he wrote, "at about the worst possible time, spent nearly five years fighting that with a determination that bordered on obsession, and turned up the hitter who could put a ball in the top deck and the outfielder whose arm was again golden. It was all so marvelous and strange."

Even stranger and more marvelous, in 2018, four years after retiring, Ankiel took the pitching mound at an exhibition game of former professional players, the first time he'd done so in public in nearly twenty years since his incident against the Braves.

He faced one batter and struck him out.

Now, to begin learning the hidden techniques for harnessing our inner voice, we need look no further than one of the more remarkable students I've ever taught. A spy from West Philadelphia.

CHAPTER THREE

Zooming Out

*H*ave you ever killed someone?" the examiner asked.

If she had been anywhere else, with anyone else, and if her future hadn't been hanging in the balance with this absurd yet apparently crucial question, she would've rolled her eyes.

"Like I told you last time," Tracey said. "No, I have never killed anyone."

Of course, I haven't, she thought. *I'm seventeen! I'm not a killer.*

This was her second polygraph with the National Security Agency, the United States' highly secretive intelligence organization. Tracey's body—her heart rate and her breathing—had betrayed her when she was asked this question the first time around, and the squiggly readout indicated that she was lying. Now, two months later, she found herself sitting in the same nondescript office in the middle of Maryland undergoing a second polygraph test.

What if they don't believe me again? she wondered, her inner voice providing an anxious running commentary as the examiner looked at her inscrutably. She knew the answer to her own question: If they didn't believe her, the future she had been dreaming of would disappear.

For as long as Tracey could remember, she knew she wanted more than the life she was born into. School and learning had always come easy to her, even if lots of other things hadn't. She grew up in a tough neighborhood in West Philadelphia, and although her family wasn't poor, money placed limitations on her future.

During her freshman year in high school, Tracey learned about a program at a boarding school in the Northeast that allowed gifted students from across the country to complete the last two years of high school on an accelerated track that set them up for success in elite colleges. While the thought of leaving her family and uprooting her existence to a new environment was daunting, the prospect of meeting new people, being challenged intellectually, and escaping the life she had known until then appealed to her. She worked hard on the application and got in.

Boarding school exposed Tracey to a new world of friends and ideas that, for the first time in her life, truly tested her. Although she sometimes felt out of place amid her mostly white peers from privileged backgrounds, she was happy.

As one of the few African American students in her program, Tracey frequently found herself invited to events to help raise money for the school. Stories like hers tended to open the wallets of wealthy donors. During one such event, she met a man named Bobby Inman, the former director of the U.S. National Security Agency.

During their conversation, Inman told her about a highly selective undergraduate training program the NSA offered to the country's most talented and patriotic students. He encouraged her to apply. She did and the NSA called her in for the interview in which she failed her first polygraph test, making her doubt that her dreams would become a reality. The second time around, however, she managed to control her nerves, and the NSA no longer suspected her of murder, if they even really had in the first place. Her life was about to change in dramatic ways, though her first polygraph experience would end up being a harbinger of things to come: the challenges of managing her inner voice.

At first blush, the terms of the scholarship were everything she wanted. The NSA would cover the entire cost of Tracey's college education and provide her with a generous monthly stipend. Of course, there were conditions. She would have to spend her summers training to become a top secret analyst and then work for the NSA for at least six years after graduating. Still, it was an incredible opportunity, especially when she got into Harvard that spring. Tracey had earned herself a free Ivy League education and a thrilling future.

A few weeks before classes at Harvard began, she got her first taste of what working with the NSA would be like. During her weeklong onboarding, she received top secret clearances allowing her to access highly classified information. She also learned about the details of the restrictions that came along with her scholarship. She could major only in a handful of subjects that were central to the NSA's interests: subjects like electrical engineering, computer science, and math. She couldn't date or maintain close friendships with students from other countries. She couldn't study abroad. She was discouraged from playing varsity sports. Slowly but surely Tracey's scholarship, her golden ticket, was morphing into a pair of golden handcuffs.

While other freshmen in her dorm mingled freely, Tracey found herself on guard. In the past, she had been the one profiled. Now at mixers she was doing the profiling, quickly scanning people's faces and vocal intonations for clues about where they came from out of fear that she might become friends with—or even worse, perhaps even feel attracted to—someone from a distant land. She felt likewise constricted by the math and engineering classes she was enrolled in, which were unlike the excitingly diverse courses so many of her peers were taking. As she hurried along the tree-lined paths of Harvard Yard between classes, her thoughts curled inward on all the not-great things about this supposedly great opportunity. She wondered if she had made a mistake.

Time passed. As she went from being a freshman to a sophomore and then a junior, Tracey felt increasingly lonely. She was drowning, as she put it, in her "inner dialogue." She couldn't talk about how she spent her summers—the training in cryptography and circuit-board building, or learning how to scale rooftops to splice antennas. But her feelings of isolation were only one source of stress. Another was the fact that engineering, one of the most challenging majors at Harvard, was proving to be the hardest academic struggle she'd ever faced, and if her grades dipped below a 3.0 GPA, she would be kicked out of the NSA program and required to pay the government back the money it had already paid her—a terrifying possibility.

The stream of her ever more negative inner voice consumed her. Ruminations about what would happen if she didn't make the grades she needed would spike before her tests. Consumed by anxiety, she began compulsively chewing on the tip of her pencil and twirling her hair during exams. Her nervous tics provided her with a strange sense of comfort. Despite her best attempts to maintain the outward appearance that everything was okay, her body disappointed her once again, in a different way than it had during her first polygraph exam. Precisely when she began to stress out about her grades, she would develop cystic acne on her face, pus-filled pimples beneath the top layer of her skin that required cortisone injections. It was as though the chatter brewing beneath her surface were too extreme to contain. She didn't know how much more she could take.

She felt as if she had two options: fail out or drop out.

Becoming a Fly on the Wall

Tracey's story, like the stories of most people whose internal conversations become pools of negativity, is an exercise in distance—the distance we do or don't have from our problems.

We can think of the mind as a lens and our inner voice as a button that zooms it either in or out. In the simplest sense, chatter is what happens when we zoom in close on something, inflaming our emotions to the exclusion of all the alternative ways of thinking about the issue that might cool us down. In other words, we lose perspective. This dramatically narrowed view of one's situation magnifies adversity and allows the negative side of the inner voice to play, enabling rumination and its companions: stress, anxiety, and depression. Of course, narrowing your attention isn't a problem in and of itself. To the contrary, it's often essential to helping us address challenging situations and the feelings that arise from them. But when we find ourselves stuck on our problems and lose the ability to flexibly zoom out—to gain perspective—that's when our inner voice turns into rumination.

When our internal conversation loses perspective and gives rise to intensely negative emotions, the brain regions involved in self-referential processing (thinking about ourselves) and generating emotional responses become activated. In other words, our stress-response hardware starts firing, releasing adrenaline and cortisol, flooding us with negative emotions, which only serve to further rev up our negative verbal stream and zoom us in more. As a result, we're unable to get a wider-angle view that might reveal more constructive ways of handling the emotionally trying situations we encounter.

But our brains evolved not just to zoom in when we confront difficulties but also to *zoom out,* though the latter is much more challenging during times of stress. The mind is flexible, if we know how to bend it. If you have a fever, you can take something to bring it down. Likewise, our mind has a psychological immune system: We can use our thoughts to change our thoughts—by adding distance.

Psychological distance, of course, doesn't eliminate a problem. If, for instance, Tracey had been able to step back from her high-pressured predicament and settle into a less anguished state, she nonetheless would still have been in debt to the NSA, with her future hanging in the balance. Similarly, even if Rick Ankiel had been able to retain his pitch, he still would've been standing on the mound pitching in the playoffs on national television. Distance doesn't solve our problems, but it increases the likelihood that we can. It unclouds our verbal stream.

The big question, then, is this: When chatter strikes, how do we gain psychological distance?

As it happened, around the same time that Tracey was sitting in her freshman dorm room at Harvard, I was three and a half hours south down the highway in Manhattan, a graduate student in psychology sitting in the basement of Columbia University's dingy Schermerhorn Hall thinking about a remarkably similar question. How can people reflect on their negative experiences, I wondered, without getting sucked down the rumination vortex? Answering that question was the reason I had decided to attend Columbia to train with my adviser, Walter Mischel, a groundbreaking scientist whom most people know as the Marshmallow Man.

Walter was akin to royalty in psychology for developing what the public now calls the marshmallow test, a paradigm for studying self-control that involved bringing kids into the lab for an experiment and presenting them with a simple choice: They could have one marshmallow now, or if they waited for an experimenter to return, they could have two. It turned out that children who waited longer ended up performing better on their SATs as teens, were healthier as they got older, and coped better with stress in adulthood than those who immediately grabbed the gooey marshmallow. But even more important than documenting these striking long-term outcomes, the so-called marshmallow test (its real name is the delay of gratification test) helped revolutionize science's understanding of the tools people have to control themselves.

By the time I arrived at Columbia, Walter and his then postdoctoral student Özlem Ayduk had already become interested in examining how to help people think about painful experiences without succumbing to chatter. At the time, one of the dominant approaches to battling inner-voice rumination was *distraction*. Several studies had shown that when people find themselves sucked into negative verbal thinking, diverting their attention away from their problems improved the way they felt. The downside of this approach, however, is that distraction constitutes a short-term fix—a Band-Aid that obscures the wound without healing it. If you go to the movies to escape the adversities of real life, your problems are still there waiting for you when you leave the theater. Out of sight, in other words, isn't actually out of mind, because the negative feelings remain, eagerly waiting to be activated again.

Oddly, at this time, the idea of distancing had fallen out of vogue in psychology. In 1970, Aaron Beck, one of the founders of cognitive therapy and an influential figure in mental health, proposed that teaching patients how to objectively scrutinize their thoughts, a process he called "distancing," was a central tool that therapists should employ with their patients. In the ensuing years, however, distancing had come to be equated with avoidance—with *not* thinking about your problems. But in my mind, there was nothing inherently avoidant about distancing. In theory, you could use your mind to frame your problems from a zoomed-out perspective.

This approach differed from the meditative practice of mindfulness in that the goal wasn't to stand apart and watch one's thoughts drift by without engaging with them. The point *was* to engage, but to do so from a distanced perspective, which isn't the same thing as an emotionally avoidant one. That was the essence of my dad's teachings and what I had spent so much time doing growing up. So, Walter, Özlem, and I began thinking about the different ways people could "step back" from their experiences to reflect on them more effectively. We landed on a tool we all possess: our ability to imaginatively *visualize*.

A powerful optical device of sorts is built into the human mind: the ability to see yourself from afar. This mental home theater, it turns out, projects scenes when we think about unpleasant experiences from the past or imagine possible anxiety-producing scenarios in the future. They are almost like videos stored on a phone. Yet these scenes aren't entirely fixed. Studies show that we don't see our memories and reveries from the same perspective every time. We can view them from different perspectives. For example, sometimes we replay a scene happening through our own eyes as though we were right back in the event in the first person. Yet we can also see ourselves *from the outside,* as if transplanted to another viewpoint. We become a fly on the wall. Could we harness this ability to better regulate our inner voice?

Özlem, Walter, and I brought participants into the lab to find out. To do so, we asked one group to replay an upsetting memory in their minds through their own eyes. We asked another group to do the same, only from a fly-on-the-wall perspective, visually observing themselves like a bystander. Then we asked the participants to work through their feelings from the perspective they had been asked to adopt. The differences in the verbal stream characterizing the two groups were striking.

The *immersers*—the people who viewed the event from a first-person perspective—got trapped in their emotions and the verbal flood they released. In their accounts describing their stream of thoughts, they tended to zero in on the hurt. "Adrenaline infused. Pissed off. Betrayed," one person wrote. "Angry. Victimized. Hurt. Shamed. Stepped-on. Shitted on. Humiliated. Abandoned. Unappreciated. Pushed. Boundaries trampled upon." Their attempts to "go inside" and work through their internal conversations just led to more negative feelings.

The fly-on-the-wall group, meanwhile, offered contrasting narratives.

Where the immersers got tangled in the emotional weeds, the *distancers* went broad, which led them to feel better. "I was able to see the argument more clearly," wrote one person. "I initially empathized better with myself but then I began to understand how my friend felt. It may have been irrational but I understand his motivation." Their thinking was clearer and more complex, and, sure enough, they seemed to view events with the insight of a third-party observer. They were able to emerge from the experience with a constructive story. The experiment provided evidence that stepping back to make sense of our experiences could be useful for changing the tone of our inner voice.

Soon after, in more studies, we and others discovered that zooming out in this way also reined in people's fight-or-flight cardiovascular response to stress, dampened emotional activity in the brain, and led people to experience less hostility and aggression when they were provoked—the kind of situation that is fertile ground for stoking chatter. We also found that this distancing technique worked not just with random collections of college students but also with those struggling with more extreme forms of inner-voice torment. For example, people with depression and even highly anxious parents grappling with their children's undergoing painful cancer treatments. Yet our findings were, at this point, still limited. They related only to how distancing affects us *in the moment*. We also wanted to know whether it would have lasting effects, shortening the amount of time people spent ruminating.

It turned out we weren't the only ones interested in exploring this question.

Not long after we published our initial work, a research team at the University of Leuven in Belgium, led by Philippe Verduyn, devised a clever set of studies to look at whether people's tendencies to distance in daily life, outside a laboratory setting, influenced how long their emotional episodes lasted.

They found that distancing by adopting an observer perspective shortened the duration of people's negative moods after they experienced events that led them to feel angry or sad. Distancing could put out chatter brush fires before they grew into longer-lasting conflagrations.

This dampening quality of distancing could, however, have an unintended consequence. Distancing shortened both negative *and* positive experiences. In other words, if you got a promotion at work and stepped back to remind yourself that status and money don't really matter in the grand scheme of things and that we all die in the end anyway, your well-deserved joy would decrease. The takeaway: If you want to hold on to positive experiences, the last thing you want to do is become a fly on the wall. In such cases, immerse away.

By this time, it had become clear that we are all prone to either psychological immersion or psychological distance when we reflect on emotional experiences, though we aren't locked into either state. The tendency we have shapes the patterns of our inner voice, but fortunately so does our ability to consciously alter our perspective.

Along with our work and Verduyn's, a slew of other studies published around the same time began to shift our understanding of the role distancing plays in helping people control their emotions. Researchers at Stanford, for example, linked adopting the perspective of a detached observer with less rumination over time. Across the Atlantic, researchers at Cambridge found that teaching people to "see the big picture" reduced intrusive thinking (the kind that drains executive functions) and avoidance of painful memories. Other experiments demonstrated that even shrinking the size of an image that causes distress in one's imagination reduces how upset people become when they view it.

Still other work applied the concept of distancing to education, showing how leading ninth graders to focus on the big-picture reasons for doing schoolwork—for instance, emphasizing how doing well in school would help them land their desired jobs and contribute to society as adults—led them to earn higher GPAs and stay more focused on boring but important tasks. Distance, then, helps us deal better not only with the big emotions we experience from upsetting situations but also with the smaller yet crucial daily emotional challenges of frustration and boredom that come with the important tedium of work and education.

All of this taught us that taking a step back could be effective for helping people manage their chatter in a variety of everyday contexts. But we would soon learn that gaining mental distance also has positive implications for something else important: wisdom.

Solomon's Paradox

It was around 1010 B.C.E. The maternal dreams of a woman in Jerusalem, named Bathsheba, finally came true. After the loss of her first child as an infant, she now gave birth to a second child: a healthy young baby boy whom she named Solomon. As the Bible tells us, this was no ordinary baby. The son of David (of Goliath fame), Solomon grew up and went on to become king of the Jewish people. A peerless leader, he was respected not only for his military might and economic acumen but also for his wisdom. People would travel from distant lands to seek his counsel.

The dispute he most famously settled was between two women who both claimed to be the mother of the same child.

He suggested that they cut the child in half, and when one of the women protested, he was able to identify her as the true mother. In an ironic twist of fate, however, when it came to Solomon's own life, he wasn't so savvy. Amorous and short-sighted, he married hundreds of women from different faiths and went to great lengths to please them, building elaborate temples and shrines so they could worship their gods. This eventually alienated him from his own God, and the people he ruled, which would finally lead to his kingdom's demise in 930 B.C.E.

The asymmetry in King Solomon's thinking is a chatter parable that embodies a fundamental feature of the human mind: We don't see ourselves with the same distance and insight with which we see others. Data shows that this goes beyond biblical allegory: We are all vulnerable to it. My colleagues and I refer to this bias as "Solomon's Paradox," though King Solomon is by no means the only sage who could lend his name to the phenomenon.

Take a little-known story about one of the wisest men in U.S. history, Abraham Lincoln, who, in 1841, was in a rut both professionally and romantically. He had yet to establish himself as a lawyer to the extent he desired. He was also anguishing over doubts about his feelings for his fiancée, Mary, because he had fallen in love with another woman. Immersed in his problems, he sank into depression, or what one historian has called "Lincoln's melancholy."

The following year, when the future president had begun to recover his hope and clarity, a good friend of his, Joshua Speed, fell into similar doubts about his own engagement. Now in a different role, Lincoln was able to offer Speed sound advice he hadn't been able to marshal with regard to his own situation. He told Speed that his ideas about love were the problem, not the woman he was engaged to marry. Lincoln later reflected, as Doris Kearns Goodwin writes in her book *Team of Rivals,* that "had he understood his own muddled courtship as well as he understood Speed's, he might have 'sailed through clear.'"

Before we look at how distancing can lead to wisdom, it's worth taking a moment to ask what wisdom actually is in practice. In a rigorous field like psychology, an amorphous-seeming concept like wisdom at first appears hard to define. Nonetheless, scientists have identified its salient features. Wisdom involves using the mind to reason constructively about a particular set of problems: those involving uncertainty. Wise forms of reasoning relate to seeing the "big picture" in several senses: recognizing the limits of one's own knowledge, becoming aware of the varied contexts of life and how they may unfold over time, acknowledging other people's viewpoints, and reconciling opposing perspectives.

Although we generally associate wisdom with advanced age, because the longer you live the more uncertainty you will have experienced and learned from, research indicates that you can teach people how to think wisely regardless of their age—through gaining distance.

Take a study that Igor Grossmann and I did in 2015. We presented people with a dilemma and asked them to predict how it would unfold in the future. One group of participants was asked to imagine their partner had cheated on them, while the other group imagined the same exact thing happening to a friend—a practical method of creating psychological distance.

While some people may understandably think that outrage is the wisest response to discovering that your partner cheated on you, our interest was in whether distance would decrease rather than increase conflict by cultivating a wise response. As we expected, people were much wiser when they imagined the problem was happening to someone else. They felt it was more important to find compromise with the person who had cheated, and they were also more open to hearing that person's perspective.

Another illustration of how people can use distance as a hatch to escape from Solomon's Paradox comes from research on medical decision making. Few contexts are more chatter provoking—and consequential—than having to make an important decision about your health. Uncertainty surrounding physical pain or illness, never mind mortality, bloats the verbal stream with worry, which can cloud our judgment and lead us to make poor decisions that, ironically, further impair our health.

In one large-scale experiment, a group of scientists gave people a choice: do nothing and have a 10 percent chance of dying from cancer, or undergo a novel treatment that has a 5 percent chance of killing you. Obviously, the second option is better, because the risk of death is 5 percent less. And yet, consistent with prior research indicating that people often choose to do nothing rather than something when it comes to their health, 40 percent of participants chose the more life-threatening option. But—and this is a big but—when the same people were asked to make this decision for someone else, only 31 percent made the bad choice. When you frame this percentage difference in terms of the number of cancer diagnoses per year—18 million—that adds up to more than 1.5 million people who could sabotage their own best course of treatment. But this lack of wisdom, brought about by a lack of mental distance, can also influence other areas of our life.

Daniel Kahneman, the Nobel Prize–winning psychologist

and author of *Thinking, Fast and Slow*, has written that one of his most informative experiences involved learning how to avoid an "inside view" and adopt an "outside view." As he frames it, an inside view limits your thinking to your circumstances. Because you don't know what you don't know, this often leads to inaccurate predictions about potential obstacles. The outside view, on the other hand, includes a broader sample of possibilities and thus more accuracy. You're able to better foresee obstacles and prepare accordingly.

Although Kahneman's views pertain to accurately predicting the future, research shows that the ability to step outside oneself—another way of saying mental distance—is helpful for decision making more generally. It can help us get past information overload—for instance, when we're evaluating contrasting features and prices while car shopping—so that we can attain clarity. It can roll back "loss aversion," the concept Kahneman popularized referring to the fact that people are much more sensitive to losses than they are to gains. Additionally, it can make people more compromising and willing to tolerate alternative views. In one study conducted right before the 2008 U.S. presidential election, Igor and I found that asking people to imagine a future in which their chosen candidate lost the election from a distanced perspective (we asked them to imagine that they were living in another country) led them to become less extreme in their political views and more open to the idea of cooperating with people who supported the opposing candidate.

These positive interpersonal and wisdom-enhancing effects of distancing make this skill invaluable to another area of life where we often experience inner-voice ranting: our romantic relationships. My colleague Özlem and I wondered how distance might factor into intimate-partner harmony. So over twenty-one days we profiled the tendency of people to distance each time they fought with their romantic partner. We found that whether people "distanced" or "immersed" when thinking about problems in their relationships influenced how they argued. When an immerser's partner argued calmly, the immerser responded the same way—with similar patience and compassion. But once their partners began to show the slightest hint of anger or disdain, the immersers responded in kind. As for the distancers, when their partners talked calmly, they too remained calm. But even if their partner got worked up, they were still able to problem solve, which eased the conflict.

A subsequent experiment took this research even further by showing that teaching couples to distance when they focused on disagreements in their relationships buffered against romantic decline. Over the course of a year, spending twenty-one minutes trying to work through their conflicts from a distanced perspective led couples to experience less unhappiness together. If not exactly a love potion, distancing does seem to keep the flame of love from being extinguished.

All this research demonstrates how useful stepping back can be for changing the nature of the conversations we have with ourselves. Yet more broadly, it also shows how we can reason wisely about the most chatter-provoking situations we face—those that involve uncertainty, which requires wisdom. But what's striking to me about all this work is that it demonstrates just how many ways there are to get psychological distance, how many options our mind gives us for gaining perspective. But sometimes we need more than wisdom. As Tracey would learn at Harvard, we need new stories—imagined narratives that also add distance—which we create by harnessing the power of the time machine in our own minds.

Time Travel and the Power of the Pen

There Tracey was, sitting in her dorm room each night, gnawing away at her pencil eraser, tormented by her acne, her inner voice spiraling into despondency with the split demands of being a covert agent in training and a lonely scholarship student at an elite university. Helplessly immersed in her anxiety, she finally spoke to therapists at Harvard and the NSA. Much to her disappointment, neither counselor really helped. She remained as alone as ever—or did she?

As a hobby, but seeming to sense it would aid her in some way, Tracey embarked on a family history project. She was fascinated by the long chain of people and events that had brought her into existence. So, during breaks from school, when she wasn't required to be at the NSA, she chased down stories from her past. Doing so led her to ride on the backs of motorcycles with relatives around Lake Michigan and walk the shores of Lake Merritt in California, wander the sticky streets of New Orleans's French Quarter with two aunts, and make grave rubbings from the family headstones that dotted the cemetery down the road from her ancestors' burned-down farm in central Texas.

As her relatives opened up to her, Tracey heard about the struggles of being part of one of the first African American families living in Kalamazoo, Michigan. She discovered that her great-grandmother had been a voodoo practitioner in a relationship with a white man, her great-grandfather, and learned about the prayers she cast to ward off evil spirits. And after careful but persistent prodding, she eventually got different relatives to talk about the most painful and oppressive chapters of her family's past in the United States. She confirmed that she was the great-great-grandchild of slaves and learned that one of her great-grandfathers had been lynched, while another had been conscripted into the Confederate army. She also discovered she was a descendant of George Washington's.

The deeper Tracey delved into her family's history, the calmer she noticed herself feeling when she returned to Harvard. On the one hand, as she tapped into the legacy of her forebears, she seemed to be demonstrating to the world that a descendant of slaves could achieve success at one of the most prestigious institutions in the world. In spite of her difficulties at Harvard, this historical perspective gave her a bird's-eye view of how far she had come, even making her think her ancestors would be proud of her. At the same time, learning about the suffering that her forebears had endured also helped her put her trials and tribulations in perspective. In her mind, the anxiety surrounding not making grades and not being able to date whom she wanted paled in comparison to the torment her ancestors must have experienced toiling away as slaves. She had become a fly on the wall not just of her own life but of generations of lives—the long line of ancestors who survived the transatlantic slave crossing and eventually flourished in the United States over time. This dramatically calmed her inner voice.

Several studies back up scientifically what Tracey experienced personally, revealing that the ability to strategically time travel in one's mind can be a tool for creating positive personal narratives that reroute negative inner dialogues. But the benefits of mental time travel aren't restricted to adopting a bird's-eye view of the past to weave together a positive story about the present. You can also benefit by mentally time traveling *into the future,* a tool called *temporal distancing.* Studies show that when people are going through a difficult experience, asking them to imagine how they'll feel about it ten years from now, rather than tomorrow, can be another remarkably effective way of putting their experience in perspective. Doing so leads people to understand that their experiences are temporary, which provides them with hope.

In a certain sense, then, what temporal distancing promotes is one of the facets of wisdom: the understanding that the world is constantly in flux and circumstances are going to change. Recognizing that feature of life when it comes to negative experiences can be tremendously alleviating. It is what helped me, for example, cope with what was arguably the most chatter-provoking event of the past century: the 2020 COVID-19 pandemic.

As schools closed, quarantines began to take effect and the world outside became quiet; chatter began to brew in my mind just like millions of other people. Will social distancing affect my children's well-being? How am I going to survive without leaving the house for weeks? Will the economy ever improve? Focusing on how I would feel once the pandemic ends made me realize that what we were going through was temporary. Just as countless pandemics had come and gone in the long history of our species, so too would the COVID-19 threat eventually pass. That buoyed my inner voice.

My colleague Özlem has found that temporal distancing helps people manage major stressors like the loss of a loved one but also more minor yet still critical ones, like looming work deadlines. And best of all, this technique doesn't just make you feel better; it even improves your love life by making relationships and arguments fare better, with less blame and more forgiveness.

Alongside her family history project, Tracey also kept a journal as her college years progressed. This, too, became a medium for gaining distance. Although journaling has surely been around nearly as long as the written word, it is only in the past few decades that research has begun to illuminate the psychological consolation it provides. Much of this work has been pioneered by the psychologist James Pennebaker (yes, he has the word "pen" in his name). Over the course of a long and distinguished career, he has shown that simply asking people to write about their most upsetting negative experiences for fifteen to twenty minutes—to create a narrative about what happened, if you will—leads them to feel better, visit the doctor less, and have healthier immune function. By focusing on our experiences from the perspective of a narrator who has to create a story, journaling creates distance from our experience. We feel less tied to it. Tracey journaled for years, and it helped her immensely.

Thanks to her inventive ability to pacify her internal dialogues, by the end of Tracey's senior year at Harvard her acne had abated, her nervous tics subsided, and her grades were stellar. She had subdued her chatter. After graduating from Harvard, she began her work for the NSA. She would spend the next eight years working on covert missions in conflict zones around the world. With hundreds of hours of advanced language training under her belt, she spoke French and Arabic fluently and blended seamlessly into her various assignments, many of which still remain confidential. The intelligence work she produced would be used to brief the highest levels of the U.S. government all the way up to the White House. In many ways, she would end up living the dynamic, cinematic life she had fantasized about when she first learned about the NSA scholarship in high school. To this day, Tracey still keeps a journal.

And she's now a professor at an Ivy League university (and no longer works for the government).

The strange thing about being a psychologist, especially one who studies how to control the inner voice, is that no matter what insights your research yields, you still can't escape being yourself. Which is to say, when I "go inside," I can still get lost, in spite of everything I know about how to distance. There's no other way of explaining what happened to me when I received the threatening letter from my stalker. I was aware of a variety of distancing tools to calm my chatter: adopting a fly-on-the-wall perspective, assuming a detached observer's perspective, imagining how I'd feel in the future, writing in a journal, and so on. And yet . . .

I was immersed.

I was all chatter.

I was living Solomon's Paradox.

All I could do was verbalize my panicked inner voice. Naturally, this created tension between my wife and me, and even her distanced perspective couldn't yank me out of my dialogue. My chatter was so intense it felt as if there were no way out—until suddenly I found the way.

I said my own name.

CHAPTER FOUR

When I Become You

*I*t was three o'clock in the morning and I sat in my pajamas, peering out the window of my home office, scrutinizing the night. I couldn't make out anything in the dark, but in my mind I saw very clearly the disturbing letter and deranged face of the person who had sent it, which I managed to concoct in my imagination with a little help from *Dexter* and the *Saw* movies.

After a long time, I turned away from the window.

Without really knowing what I was doing, I wandered over to my desk, sat down, and opened my computer. Somehow, even in the depth of my fear, I realized that this couldn't go on. The lack of sleep was draining me, I wasn't eating, and I was having trouble focusing at work. In this bleary-eyed state, I went "inside" again as intently as I could to find a way out of this mess. Introspection hadn't yielded much in the previous days, but I focused my mind on the problem. *What about a body-*

guard? I thought to myself. *One who specializes in protecting professors.*

As ridiculous as this sounds to me in retrospect, at the time it didn't seem ridiculous at all. But as I readied my fingers to start googling for bodyguards specially trained in defending frightened academics in the Midwest, something happened. I stopped, leaned back from my computer, and said to myself in my mind, *Ethan, what are you doing? This is crazy!*

Then something strange happened: Saying my own name in my head, addressing myself as if I were speaking to someone else, allowed me to immediately step back. Suddenly I was able to focus on my predicament more objectively. The notion that a cottage industry had developed for protecting professors with Navy SEAL–credentialed bodyguards, an idea that moments ago had seemed reasonable enough to google, now became apparent for what it really was: *lunacy.*

Once I had this realization, others quickly followed. *How is pacing the house with a baseball bat going to help?* I thought. *You have a state-of-the-art alarm system. Nothing else disconcerting has happened since you first received the letter. It was probably just a hoax. So, what are you worried about? Enjoy your life the way you used to. Think about your family, students, and research. Plenty of people receive threats that amount to nothing. You've managed worse situations. You can deal with this.*

Ethan, I said to myself. *Go to bed.*

As these thoughts began to spread like a salve on an open wound, I walked from my office to my bedroom. My heartbeat slowed, and the weight of my emotions changed. I felt lighter. And when I quietly got into bed next to my wife, I was able to do something that I had desperately wanted to do since I first received the letter: I closed my eyes without clenching my teeth, without booby-trapping the door to my bedroom, without clutching my Little League baseball bat, and I slept deeply until morning.

Saying my own name had saved me. Not from my hostile stalker, but from myself.

During the days and then weeks following that night, I kept thinking about what had happened. On the one hand, there was the uncomfortable irony that I was a psychologist who specializes in self-control and yet I had lost my self-control, never mind my rationality, albeit briefly. On the other hand, there was the scientifically intriguing observation that I had somehow regained control of my emotions and internal conversation by talking to myself as if I were another person. Normally, using one's own name is associated with eccentricity, narcissism, or sometimes mental illness, but I didn't identify with any of these. For me, at least in that moment of crisis, I had somehow managed to subdue my inner voice . . . with my inner voice.

And I had done so without even meaning to.

There's a classic finding in psychology called the frequency illusion. It describes the common experience of, say, learning a new word and then suddenly seeing it seemingly everywhere you look. In reality, the word—or whatever recent new observation you've had—has always been present in your environment with an ordinary frequency; your brain just wasn't sensitized to it before, so this creates a mental illusion.

Something similar happened to me after realizing that I had spoken to myself during a moment of tremendous emotional stress. The pattern recognition software in my mind for people talking to themselves as if they were communicating with someone else—using their names and other non-first-person pronouns—was activated. Over the next few months, then years, I noticed more and more noteworthy instances of it in several different contexts.

The threatening letter arrived in the spring of 2011, but the first case that caught my attention was actually a recollection I had of the basketball superstar LeBron James from the summer of 2010. As a lifelong Knicks fan, I had been holding out the naive hope that he would come to New York to redeem my floundering team. Instead, he appeared on ESPN to announce that he was leaving the Cleveland Cavaliers, the hometown team that had nurtured his career from its inception, to play for the Miami Heat—a high-stakes and, by his own admission, difficult decision. "One thing that I didn't want to do was make an emotional decision," LeBron explained to the ESPN commentator Michael Wilbon. A split second later, right after he articulated his goal to avoid making an emotional decision, he switched from talking about himself in the first person to talking about himself using his own name: "And I wanted to do what was best for LeBron James and what LeBron James is going to do to make him happy."

A few years later, I came across a video of the future Nobel Peace Prize winner Malala Yousafzai on *The Daily Show with Jon Stewart*. In the summer of 2012, fourteen-year-old Malala was living in the Swat valley of Pakistan with her family when she received arguably one of the most stressful pieces of news imaginable: The Taliban had vowed to assassinate her as punishment for her outspoken advocacy of girls' rights to education. When Stewart asked her how she responded to learning of the threat against her, Malala inadvertently revealed that employing her own name to coach herself had been key. After beginning to recount her experience in the first person, as she narrated the story and arrived at its most fearsome moment, she told Stewart, "I asked myself, 'What would you do, Malala?' Then I would reply to myself, 'Malala, just take a shoe and hit him' " . . . But then I said, 'If you hit a Talib with your shoe, then there will be no difference between you and the Talib.' "

The examples kept cropping up, not just in pop culture contexts—such as the actress Jennifer Lawrence pausing during an emotional interview with a *New York Times* reporter to say to herself, "O.K., get ahold of yourself, Jennifer"—but also in historical instances that had been hiding in plain sight. There was already a term for talking about oneself in the third person, "illeism," which was frequently used to describe the literary device Julius Caesar had employed to narrate his work on the Gallic Wars, in which he had participated. He wrote about himself by using his own name and the pronoun "he" instead of the word "I." And then there was the American historian Henry Adams's Pulitzer Prize–winning autobiography, published in 1918, which he narrated entirely in the third person. In keeping with this stylistic approach, he didn't title the book *My Education* or something similar. He called it *The Education of Henry Adams.*

By this time, I had already shared my observations about how people use their own names and second- and third-person pronouns to talk to themselves with my students and colleagues. As a result, a conversation had gotten under way in the lab, and we had begun to examine the relationship between language and distance. We had a strong intuition that using one's own name—silently in one's own head, that is, not talking to oneself aloud in ways that elicit raised eyebrows and disrupt social norms—was a tool that helped people control their inner voice.

Of course, all of the "evidence" I had come across was anecdotal. It wasn't scientific proof of anything, though it did seem to suggest a common pattern in human behavior. For years my colleagues and I had been studying approaches to distancing, yet all the techniques we had uncovered required both time and concentration, whereas using one's name to mentally speak to oneself in a moment of distress had taken neither. Could talking to yourself as if you were someone else be its own form of distancing?

Say Your Name

"Are you serious?" the participant in our experiment asked.

"Yes," the experimenter told him. "Follow me."

She led him down the hallway.

Like the other volunteers who had come into our lab, he had known only that he was going to participate in an experiment on language and emotion. What none of the volunteers knew until they arrived for the study was the method we would be using, one of the most powerful techniques scientists have at our disposal for stressing people out in the lab: We asked them to engage in public speaking in front of an audience without giving them sufficient time to prepare. In doing so, we hoped to gain a better understanding of how silently referring to ourselves using our own names (and other non-first-person pronouns like "you") might help people control an inner voice agitated by circumstances like the ones we had concocted.

When they arrived, we told the volunteers that they would have to deliver a five-minute speech to a group on why they were qualified to land their dream job. Then we escorted them into a small windowless room, where they had five minutes to prepare their presentations without being able to take any notes. Our idea was that if we asked some participants to use non-first-person language while thinking to themselves before the speech, they would have more mental distance, which would help them manage their nerves.

Our theory wasn't based only on my experience or the words of Malala, LeBron James, and others. Previous research had indicated that a high usage of first-person-singular pronouns, a phenomenon called I-talk, is a reliable marker of negative emotion. For example, one large study performed in six labs across two countries with close to five thousand participants revealed a robust positive link between I-talk and negative emotion. Another study showed that you can predict future occurrences of depression in people's medical records by computing the amount of I-talk in their Facebook posts. All of which is to say, talking to oneself using first-person-singular pronouns like "I," "me," and "my" can be a form of linguistic immersion.

A natural question arose: What would happen if you not only reduced a person's tendency to think about themselves in the first person but actually had them refer to themselves as if they were interacting with someone else? Our idea was that using your own name, while also employing the second and third person, created emotional distance because it makes you feel as if you were talking to another person when you're talking to yourself. For example, rather than thinking to oneself, *Why did I blow up at my co-worker today?* a person could think, *Why did Ethan blow up at his co-worker today?*

After the five-minute speech preparation period was over, we randomly divided the participants into two groups: one in which they reflected on their anxieties surrounding their upcoming speech using the first-person pronoun "I"; and the other in which they did the same but only using non-first-person pronouns and their own name. After they were done, we took them down the hall to deliver their presentations in front of a panel of judges who were trained to maintain stoic facial expressions and a large video camera that was distractingly positioned right in front of them. It was showtime.

As we predicted, participants who used distanced self-talk reported that they experienced less shame and embarrassment after giving their speech compared with participants who used immersed self-talk. They also ruminated less about their performance afterward. In their descriptions of their mental experiences, instead of highlighting their nervousness or the difficulty of the task, they said that their inner voices focused on the fact that nothing of real consequence was actually at stake.

Remarkably, as we coded the videos and dug deeper into the data from the experiment, it wasn't just the participants' emotional responses that differed. Judges who watched videos of participant's speeches indicated that people in the distanced self-talk group performed better on the task as well.

We had uncovered a novel distancing tool hidden in the mind: *distanced self-talk.* As our experiments and others later demonstrated, shifting from the first-person "I" to the second-person "you" or third-person "he" or "she" provides a mechanism for gaining emotional distance. Distanced self-talk, then, is a psychological hack embedded in the fabric of human language. And we now know that its benefits are diverse.

Other experiments have shown that distanced self-talk allows people to make better first impressions, improves performance on stressful problem-solving tasks, and facilitates wise reasoning, just as fly-on-the-wall distancing strategies do. It also promotes rational thinking. For instance, during the height of the 2014 Ebola crisis, some people were terrified about contagion in the United States. So we ran a study over the internet with people across the United States. We found that people who were anxious about Ebola and were asked to switch away from using "I" to using their own names to reflect on how the Ebola scare would play out in the future found more fact-based reasons not to worry, which predicted a decrease in their anxiety and risk perception. They no longer thought it was so likely that they would contract the disease, which was both a more accurate reflection of reality and a muzzle on their previously panicked inner voices.

Research also shows that distanced self-talk can have implications for helping people deal with one of the most chatter-provoking scenarios I've studied: having to choose between our love of others we care for and our moral principles. For instance, a person we know commits a crime, and we're forced to decide whether to protect or punish them. Studies show that when this internal conflict occurs, people are considerably more likely to protect those they know rather than report them, a phenomenon that we see define decisions in everyday life time and again—for example, the university administrators and gymnastics officials who failed to stop the now-convicted child-molesting physician Larry Nassar.

If the reason why we are motivated to protect certain people is that we are so close to them, then it would follow that engaging in distanced self-talk should reduce these protective tendencies by allowing us to step back from ourselves and the relationships we share with others. Sure enough, across several experiments, this is exactly what we found. For example, in one study, my students and I asked people to vividly imagine observing a loved one commit a crime, like secretly using another person's credit card, and then being approached by a police officer who asks if they saw anything. Participants who reflected on what they should do using their own name (for example, *What facts is Maria considering when making this decision?*) were more likely to report severe offenses to the police officer.

While these findings demonstrated the power of distanced self-talk, they didn't explore another property that makes it so valuable: its speed. One of the things I found most interesting about saying my own name to calm myself down was how remarkably easy it was. Normally, it takes time to regulate our emotions. Just think of the effort involved in mentally traveling through time to imagine how you'll feel differently about something in the future, or writing a journal entry to contemplate your thoughts and feelings, or even closing your eyes to picture a past experience from a fly-on-the-wall perspective. These are all empirically validated self-distancing tools. Yet, because of the effort they require, they're not always easy to implement in the heat of the moment.

Now think about my experience. All I did was say my name, and it put my inner voice on a totally different trajectory, almost like switching the direction a train goes when it comes to a Y-juncture. Distanced self-talk appeared to be quick and powerful, unlike so many other emotion-regulation strategies. How could that be?

In linguistics, "shifters" refer to words, like personal pronouns (such as "I" and "you"), whose meaning changes depending on who is speaking. For example, if Dani asks, "Can *you* pass me the ketchup?" and Maya answers, "Sure, here *you* go," the person that "you" addresses changes. It refers to Maya initially but then Dani. Most children figure out that language works this way by the time they are two years old and can switch perspectives in this way incredibly fast, within milliseconds.

The concept of shifters demonstrates how powerful certain words can be for switching our perspectives. Our idea was that distanced self-talk might operate through a similar mechanism, producing a virtual automatic shift in perspective requiring minimal effort. Using this lens on language and psychological distance, the Michigan State University psychologist Jason Moser and I designed an experiment to measure how quickly distanced self-talk works. But instead of listening to people's inner voices, we looked at their brains.

In our experiment, we asked participants to think about how they felt each time they saw a disturbing photograph, using either immersed language (*What am I feeling?*) or distanced language (*What is Jason feeling?*). As they did this, we monitored the electrical activity of their brains using an electroencephalogram machine, which provides a useful means of determining just how quickly different psychological operations work in the brain.

The results indicated that participants displayed much less emotional activity in the brain when they used distanced language to reflect on their feelings after viewing the disturbing pictures. But the crucial finding was how long it took the participants to feel the relief of distance. We saw changes in emotional activity emerge within one second of having people view a negative picture.

One tiny second. That was it.

Equally exciting to us, we didn't find evidence to suggest that this kind of self-talk overtaxed people's executive functions. This was crucial, because more effortful distancing techniques create a Catch-22 of sorts: When our chatter is buzzing, it drains us of the neural resources we need to focus, get distance, and regain control of our inner voice. Yet distanced self-talk sidesteps this conundrum. It is high on results and low on effort.

If changing the words we use to think about ourselves offers a hyper-speed form of distancing for dealing with stress, it stood to reason that it should also influence the stream of our inner voice. As it turns out, distanced self-talk can do this by harnessing a capacity we all possess: the ability to interpret sources of stress as challenges rather than threats. To see how this works, let's drop in on an old neighbor.

Get to It, Fred

If you grew up or had children in the United States between 1968 and 2001, you probably recall Fred Rogers's soothing voice on his legendary thirty-minute television program, *Mister Rogers' Neighborhood*. But beneath his serene persona, Rogers's inner voice could torment him, just like the rest of us. We know this because his inner critic is on full display in a letter he typed to himself in 1979, shortly after returning from a three-year break from doing his show:

> Am I kidding myself that I'm able to write a script again? Am I really just whistling Dixie? I wonder. If I don't get down to it I'll never really know. Why dan't . . . I trust myself. Really that's what it's all about . . . that and not wanting to go through the agony of creation. AFTER ALL THESE YEARS IT'S JUST AS BAD AS EVER. I wonder if every creative artist goes through the tortures of the damned trying to create.?. Oh, well,

the hour commeth [*sic*] and now IS when I've got to do it. GET TO IT, FRED. GET TO IT.

Rogers's strikingly vulnerable letter provides us with a raw chatter artifact of sorts, a front-row seat to observe his shifting inner voice.

The first three-quarters of the letter presents an inner dialogue that is filled with self-doubt, self-criticism, and even despair. But as the note to himself progresses, you can see Rogers building toward another way of thinking about his situation. His inner critic begins to fade out as he recognizes that regardless of his insecurities he has to deal with the task at hand— "the hour commeth . . . I've got to do it." And then he does it. He switches into using distanced language—using his own name—to convey to himself that he can in fact write his show. And with that shift in perspective he got back to work for another twenty-two years while at the same time illuminating the fork in the road we all face when confronting an overwhelming situation.

Psychologists have shown that when you place people in stressful situations, one of the first things they do is ask themselves (usually subconsciously) two questions: *What is required of me in these circumstances, and do I have the personal resources to cope with what's required?* If we scan the situation and conclude that we don't have the wherewithal needed to handle things, that leads us to appraise the stress as a *threat.* If, on the other hand, we appraise the situation and determine that we have what it takes to respond adequately, then we think of it as a *challenge.* Which way we choose to talk about the predicament to ourselves makes all the difference for our inner voice. And unsurprisingly, the more constructive framing of a challenge leads to more positive results. In Mr. Rogers's case, it allowed him to acknowledge the difficulty of creation, and then keep creating.

Several studies back up what Mr. Rogers's letter embodies. From taking math exams to performing in pressure-filled situations to coping with the toxic effects of stereotyping, people think, feel, and perform better when they frame the stressor at hand as a challenge rather than a threat. But as Mr. Rogers's use of his own name to motivate himself suggests, distanced self-talk can be the pivotal shove that sends you down the path of the challenge mindset.

Research shows that distanced self-talk leads people to consider stressful situations in more challenge-oriented terms, allowing them to provide encouraging, "you can do it" advice to themselves, rather than catastrophizing the situation. In one study that my collaborators and I performed, for example, we asked people to write about their deepest thoughts and feelings concerning an upcoming stressful event using immersed or distanced self-talk. Seventy-five percent of participants whose essays revealed the highest levels of challenge-oriented thinking were in the distanced self-talk group. In stark contrast, 67 percent of participants whose essays revealed the highest levels of threat-oriented thinking were in the immersed self-talk group.

To see how this actually played out inside participants' heads, consider what one person in the immersed group wrote: *I am afraid that I won't get a job if I mess up during an interview. And I always mess up in some way. I never know what to say, and I am always incredibly nervous. I end up in a feedback loop of nervousness causing bad interviews causing nervousness. Even if I got a job, I think I would still be afraid of interviews.*

Meanwhile, the distanced-language group's inner voices were notably different. One participant, reflecting on the insecurity he was feeling in anticipation of a date, wrote, *Aaron, you need to slow down. It's a date; everyone gets nervous. Oh jeez, why did you say that? You need to pull it back. Come on man, pull it together. You can do this.*

You don't, however, solely need to scrutinize the content of people's thoughts to see how language influences our tendency to perceive experiences as challenges or threats: You can see it in people's bodies as well. The psychological experience of challenge and threat have unique biological signatures. When you put a person in a threatening state, their heart starts pumping blood faster throughout his body. The same is true of a challenge. A key difference between the two states is how the tangle of arteries and veins that carry blood in the body responds. When a person is in a threat state, their vasculature constricts, leaving less room for their blood to flow, which over time can lead to burst blood vessels and heart attacks. In contrast, when people are in challenge mode, their vasculature relaxes, allowing blood to move easily throughout the body.

Lindsey Streamer, Mark Seery, and their colleagues at the University at Buffalo wanted to know whether distanced self-talk would lead to shifts like this in the way people's cardiovascular systems functioned. Put more simply, through distanced self-talk, could you persuade not only your mind but your body to see a situation as challenging rather than threatening? Sure enough, participants who were asked to use their name to reflect on stress before giving a public speech displayed a challenge-mode cardiovascular response. People in the immersed-language group displayed a textbook biological threat response.

If distanced self-talk can help adults, it's natural to wonder if it can benefit children as well. One of the great tasks of being a parent is teaching your children how to persevere in situations that are difficult but important, such as finding ways to help them study. With this question in mind, the psychologists Stephanie Carlson and Rachel White discovered what is known as the Batman Effect.

In one experiment, they had a group of children pretend they were a superhero as they performed a boring task designed to simulate the experience of having to complete a tedious homework assignment. The kids were asked to assume the role of the character and then ask themselves how they were performing on the task using the character's name. For example, a girl in the study who was pretending to be Dora the Explorer was instructed to ask herself, "Is Dora working hard?" during the study. Carlson and White found that the kids who did this persevered longer than children who reflected on their experience the normal way using "I." (Kids in a third group who used their own names also outperformed the I-group.)

Taking this phenomenon into even more stressful circumstances, other research with kids has linked distanced self-talk with healthy coping following the loss of a parent. For example, one child said, "No matter what, their dad loved them, and they have to think of the good things that happened . . . they can hold on to the good memories and just let the bad ones go." Conversely, children who employed more immersed language had higher incidences of post-traumatic stress symptoms and more avoidant, unhealthy coping. One child heartbreakingly said, "I still picture it—how he looked at the end. I wish he didn't have to be in pain. I'm upset that he died that way."

All of these findings highlight how a small shift in the words we use to refer to ourselves during introspection can influence our ability to control chatter in a variety of domains. Given the benefits associated with this tool, it's worth asking whether other types of distanced self-talk exist that are similarly effective in helping people manage their emotions. My colleagues and I would discover that such additional shifts exist, but their use is so subtle, pervasive, and seamless you could almost fail to notice them.

The Universal "You"

Although the chatter I experienced after receiving my letter felt unbearable before I said my own name to myself, there was one moment that brought me a sliver of relief, if only temporarily: when the police officer I met with told me that such threats were in fact a common occurrence for people with public-facing careers and they almost always blow over without incident. Plunged as I was in deep threat thinking—the letter did *not* feel like an exciting challenge—this information didn't erase my fears. But it did provide a beam of hope.

It made me feel less alone.

There is a potent psychological comfort that comes from *normalizing* experiences, from knowing that what you're experiencing isn't unique to you, but rather something everyone experiences—that, unpleasant as it is, it's just the stuff of life. When we are going through grief, relationship turbulence, professional setbacks, parenting struggles, or other types of adversity, we often feel agonizingly alone, zoomed in as we are on our problems. Yet when we talk with others and learn that they have faced similar challenges, we realize that as hard as the experience is, it happens to other people, which gives an immediate sense of perspective. *If other people got through this hardship,* our internal dialogue now reasons with us, *then so can I.* What felt extraordinary, it turns out, is in fact ordinary. This offers relief.

Now what if, instead of normalizing our experiences

through hearing other people talk about overcoming adversity or benefiting from their expertise, we could find a form of distanced self-talk with the same effect? What I mean is, could there be something built into the very structure of language that helps us think about our own personal experiences in more universal terms?

In May 2015, David Goldberg, the Silicon Valley entrepreneur and husband of Facebook's COO, Sheryl Sandberg, had an accident on a treadmill while on vacation in Mexico and died tragically. In the aftermath, Sandberg was devastated. Her life with Goldberg had disappeared, as if her future had been ripped out of her hands. In the wake of his death, she looked for ways to withstand the fierce tide of grief that threatened to suck her under. She began journaling about what she was going through—an understandable choice because, as we know, expressive writing is an effective means of gaining helpful emotional distance. Yet with the words she used in at least one entry—which she decided to publish on Facebook—she also did something curious. Notice the exact words in Sandberg's post (my italics),

> I think when tragedy occurs, it presents a choice. *You* can give in to the void, the emptiness that fills *your* heart, *your* lungs, constricts *your* ability to think or even breathe. Or *you* can try to find meaning.

At first glance, her repetitive use of the second-person "you" and "your" might seem odd. She's writing about one of the most painful personal experiences imaginable without using the most natural word for recounting her own experience: "I." Instead, she relies on the word "you," but not in the sense we've previously discussed, as if addressing herself directly like she were talking to someone else. She's using the word instead to invoke the universal nature of her hardship. It's as if she were saying, "*Anyone* can give in to the void, the emptiness that fills *everyone's* heart, *everyone's* lungs, constricts *everyone's* ability to think or even breathe. Or *anyone* can try to find meaning."

Sandberg is by no means alone in using the word "you" this way. If we look around, we can find similar usages—in everyday speech, on talk shows and radio, in song lyrics. Indeed, once you notice this phenomenon, it's hard to read interviews with athletes talking about bad games or politicians doing interviews about obstacles without noticing their use of "you" in this fashion to frame their experience more broadly.

The question, of course, is why we do this. Why do we use a word that is typically used to refer to someone else—you—to talk about our own deeply emotional experiences? My colleagues Susan Gelman and Ariana Orvell and I call this specific usage "generic 'you'" or "universal 'you.'" And we've found that it is another type of linguistic hack that promotes psychological distance.

The first thing we know about the universal "you" is that people use it to talk about norms that apply to everyone, not personal preferences. For instance, if a child holds up a pencil and asks, "What do you do with this?" an adult will typically respond, "You write with it" (not "I write with it"). In contrast, if that same child holds up a pencil and asks, "What do you like to do with this?" an adult will typically respond in a personal first-person fashion, saying, "I write with it." In other words, the generic usage of the word "you" allows us to talk about how things function generally, not our specific idiosyncratic proclivities.

We also know people use the universal "you" to make sense

of negative experiences, to think about difficult events as not unique to the self but instead characteristic of life in general, as Sandberg did in her Facebook post. For example, in one study we instructed people to either relive a negative experience or think about the lessons they could learn from the event. Participants were almost five times more likely to use the universal "you" when they were trying to learn from their negative experience than when they simply rehashed what happened. It connected their personal adversity more generally to how the world works. Participants who were asked to learn from their experience wrote statements like "When you take a step back and cool off, sometimes we see things from a different perspective," and "You can actually learn a lot from others who see things differently than you."

These kinds of normalizations provide us with the perspective we lack when mired in chatter. They help us learn lessons from our experiences that contribute to us feeling better. In other words, our use of the universal "you" in speech isn't arbitrary. It's one more emotion-management gadget that human language provides.

So, what happened after I talked to myself and fell asleep?

The next morning, I woke up and life was back to normal. I chatted over breakfast with my wife about what she had planned for the day, played with my daughter before leaving for work, and got back to all the students and research that I had neglected over the past three days. Distanced self-talk had transformed my ability to manage my chatter. And, as if my tormentor saw that he or she could no longer upset me, the letter writer never bothered me again. And yet a troubling thought stayed with me.

I had spoken to numerous people after receiving the letter, when I was at the height of my rumination. I reached out for help. And without exception, the conversations I had with friends, family, and colleagues made me feel supported. But they didn't make me feel better about the situation. They didn't soothe my inner voice the way distanced self-talk had.

The reason for this discrepancy brings us to another one of the great mysteries of the human mind. Just like the inner voice itself, other people can be a tremendous asset, but more often than we realize, they can be a liability too.

CHAPTER FIVE

The Power and Peril of Other People

*T*ragedy arrived swiftly and without warning on the campus of Northern Illinois University on a Thursday in February 2008, when a twenty-seven-year-old with a history of mental illness named Steven Kazmierczak kicked open the door to a lecture hall where a geology class was in session. Armed with a shotgun and three handguns, he stepped onto the stage that the professor was lecturing from. The 119 students sitting in the class watched in confusion, then disbelief, then terror, as the unexpected guest fired a shotgun at them, followed by another blast at their professor. Then he opened fire on them again. After discharging more than fifty rounds from different guns, he concluded his rampage by turning one of them on himself and taking his own life. Minutes later, the police descended on the gruesome scene. Twenty-one people were injured and five dead, not including Kazmierczak. The university and the small city of DeKalb, where it is located, were devastated.

After the tragedy, the community held public vigils, but many students chose to express their feelings online, posting on Facebook and memorial websites and using chat messaging programs to talk about what had happened.

One hundred and seventy miles south of DeKalb, at the University of Illinois Urbana-Champaign, the psychologists Amanda Vicary and R. Chris Fraley saw the Northern Illinois tragedy as a heartbreaking but valuable opportunity to further a line of research they were already pursuing to better understand grief and emotional sharing in real time. In science we sometimes need to look at the most painful experiences people endure to learn something valuable about how to help people navigate such events. To do so takes both delicacy and compassion, as well as commitment to the scientific method and its potential to yield insights that benefit the greater good. This was the task that Vicary and Fraley set themselves to in the aftermath of the shooting in DeKalb.

They started by emailing a large number of students from Northern Illinois to participate in a study to track how they were coping. Ten months earlier, a gunman had gone on an even more destructive rampage at Virginia Tech, killing thirty-two people and similarly leaving behind a grief-stricken community. Vicary and Fraley had also reached out to a group of Virginia Tech students shortly after that attack. Now they had two samples to pool together to get a picture of how the survivors recovered from the resulting welter of emotions.

Two weeks after the shootings, roughly three-quarters of the students in the two samples displayed symptoms of depression or post-traumatic stress. This was to be expected. Most of them were grappling with the most disturbing experience of their lives. Tragedies of the sort that the students had gone through in both Illinois and Virginia challenge a person's worldview. When that happens, some people try to avoid focusing on their traumatic memories to blunt the pain. But others actively try to make sense of their feelings, and a principal way of doing so is by communicating with others, which is what the students did. Eighty-nine percent of them joined a Facebook group to talk and read about what happened. Seventy-eight percent, meanwhile, chatted online about it, and 74 percent texted about it using their cellphones.

Most of the students found this way of releasing their chatter comforting. It allowed them to express their thoughts and feelings with others who were dealing with a similar experience, which can be a valuable form of normalization. As one Virginia Tech student said, "When I have a bout of loneliness, I can log on to Facebook or send someone an IM and I'll feel just a little more connected to people."

None of this was particularly surprising. As we already know, people are naturally disposed to sharing their thoughts with others when they are struggling with chatter, and social media and other forms of virtual connectivity provide convenient avenues for doing so. What was surprising was what Vicary and Fraley discovered when the study ended two months after the shootings.

While the students at Virginia Tech and Northern Illinois University thought that expressing their emotions to others made them feel better, the degree to which they shared their emotions didn't actually influence their depression and post-traumatic stress symptoms.

All that emoting, writing, connecting, and remembering—it hadn't been beneficial.

From Aristotle to Freud

The same year that the Northern Illinois massacre occurred, a related study was published that examined the emotional resiliency of a nationally representative sample of people living in the United States in the wake of the September 11 attacks. The researchers examined whether more than two thousand people living across the country chose to express their feelings about 9/11 during the ten days following the fall of the Twin Towers. Then they tracked the participants' physical and mental health over the next two years. The terrain of human behavior they were looking at was complicated, but their question was simple: Does sharing emotions impact how we feel over time?

What they found was remarkably consistent with what Vicary and Fraley discovered.

The people who shared their thoughts and feelings about 9/11 right after it happened didn't feel better. In fact, on the whole, they fared *worse* than the people in the study who didn't open up about how they felt. They experienced more chatter and engaged in more avoidant coping. Moreover, among those who did choose to express their feelings, the people who shared the most had the highest levels of general distress and worst physical health.

Once again, sharing emotions didn't help. In this case, it hurt.

Of course, both the college shootings and 9/11 were rare acts of extraordinary violence, which might lead you to think that sharing emotions with others is only unhelpful in the wake of tragic events. That brings us back to the work of the Belgian psychologist Bernard Rimé.

Recall the fundamental pattern in human behavior that

Rimé uncovered. When people are upset, they are strongly driven to share their feelings with others; emotions act like jet fuel that propels us to talk to other people about the thoughts and feelings streaming through our heads. But alongside this discovery, he uncovered something equally important—and certainly more surprising—which confirms that these studies on the emotional fallout of major tragedies aren't isolated cases.

In study after study, Rimé found that talking to others about our negative experiences doesn't help us recover in any meaningful way. Sure, sharing our emotions with others makes us feel closer to and more supported by the people we open up to. But the ways most of us commonly talk and listen to each other do little to reduce our chatter. Quite frequently, they exacerbate it.

Rimé's finding, along with many others, clashes dramatically with conventional wisdom. *Talking,* we are often told by popular culture, *makes you feel better.* Much self-help literature tells us this, as do many of the people around us. We hear that venting our emotions is healthy and supporting others is indispensable. It's not that simple, though there are reasons it might seem that it is.

The idea that talking about negative emotions with others is good for us isn't a recent development. It has been a part of Western culture for more than two thousand years. One of the earliest proponents of this approach was Aristotle, who suggested that people need to purge themselves of their emotions after watching a tragic event, a process he called catharsis. But this practice didn't really gain traction more broadly until two millennia later. As modern psychology was bursting to life in Europe in the late 1890s, Sigmund Freud and his mentor Josef Breuer picked up Aristotle's thread and argued that the path to a sound mind required people to bring the dark pain of their inner lives into the light. You can think of this as the hydraulic model of emotion: Strong feelings need to be released like the steam escaping from a boiling kettle.

While these cultural trappings urge us from a young age to talk to others about our feelings, the underlying drive to air our inner voice is actually implanted in our minds at an even earlier stage of our development—when we are drooling, screaming babies.

As newborns, helpless to care for ourselves or manage our emotions, we signal our distress to our caretakers, usually by wailing like little banshees (or at least my daughters did). After we get our needs met and the feeling of threat passes, our physiological arousal levels return to normal. Engaging in this process establishes an attachment to the caretaker, who often talks to the baby even before the infant can understand words.

Over time, our rapidly developing brains acquire language and soak up what our caregivers tell us about cause and effect, how to remedy our problems and deal with our emotions. This not only provides us with useful information for managing how we feel; it also provides us with the storytelling tools we need to talk to others about our experiences. This is one explanation for why communication is so entwined with chatter and why chatter is so entwined with seeking out other people.

Fortunately, there is a reason why the support we get from others so often backfires and a way to circumvent this phenomenon. Other people can be an invaluable tool for helping us subdue our chatter, and we can likewise help others with theirs. But as with any tool, to benefit from it we need to know how to properly use it, and in the case of giving and receiving support, that knowledge begins with understanding two basic needs that all humans have.

The Co-rumination Trap

When we're upset and feel vulnerable or hurt or overwhelmed, we want to vent our emotions and feel consoled, validated, and understood. This provides an immediate sense of security and connection and feeds the basic need we have to belong. As a result, the first thing we usually seek out in others when our inner voice gets swamped in negativity is a fulfillment of our *emotional* needs.

We often think of fight or flight as the main defensive reaction human beings turn to when faced with a threat. When under stress, we flee or hunker down for the impending battle. While this reaction does characterize a pervasive human tendency, researchers have documented another stress-response system that many people engage in when under threat: a "tend and befriend" response. They seek out other people for support and care.

From an evolutionary perspective, the value of this approach comes from the fact that two people are more likely to ward off a predator than one; banding together during times of need can have a concrete advantage. Supporting this idea, research indicates that affiliating with others while under stress provides us with a sense of security and connection. It triggers a cascade of stress-attenuating biochemical reactions—involving naturally produced opioids as well as oxytocin, the so-called cuddle hormone—and feeds the basic need humans have to belong. And of course, a principal way we do this is by talking. Through active listening and displays of empathy, those who counsel us on our chatter can address these needs. Satisfying them can feel good in the moment, offering one sort of relief. But this is just one-half of the equation. That is because we also need to satisfy our *cognitive* needs.

When we're dealing with chatter, we confront a riddle that demands solving. Inhibited by our inner voice run amok, we at times need outside help to work through the problem at hand, see the bigger picture, and decide on the most constructive course of action. All of this can't be addressed solely by the caring presence and listening ear of a supportive person. We often need others to help us distance, normalize, and change the way we're thinking about the experiences we're going through. By doing so, we allow our emotions to cool down, pulling us out of dead-end rumination and aiding us in redirecting our verbal stream.

Yet this is why talking about emotions so often backfires, in spite of its enormous potential to help. When our minds are bathed in chatter, we display a strong bias toward satisfying our emotional needs over our cognitive ones. In other words, when we're upset, we tend to overfocus on receiving empathy rather than finding practical solutions.

This dilemma is compounded by a commensurate problem on the helper side of the equation: The people we seek out for help respond in kind, prioritizing our emotional needs over our cognitive ones. They see our pain and first and foremost strive to provide us with love and validation. This is natural, a gesture of caring, and sometimes even useful in the short term. But even if we do signal that we want more cognitive assistance, research demonstrates that our interlocutors tend to miss these cues. One set of experiments demonstrated that even when support providers are explicitly asked to provide advice to address cognitive needs, they still believe it is more important to address people's emotional needs. And it turns out that our attempts to satisfy those emotional needs often end up backfiring in ways that lead our friends to feel worse.

Here's how talking goes wrong.

To demonstrate that they are there to offer emotional support, people are usually motivated to find out exactly what happened to upset us—the who-what-when-where-why of the problem. They ask us to relate what we felt and tell them in detail what occurred. And though they may nod and communicate empathy when we narrate what happened, this commonly results in leading us to relive the very feelings and experiences that have driven us to seek out support in the first place, a phenomenon called co-rumination.

Co-rumination is the crucial juncture where support subtly becomes egging on. People who care about us prompt us to talk more about our negative experience, which leads us to become more upset, which then leads them to ask still more questions. A vicious cycle ensues, one that is all too easy to get sucked into, especially because it is driven by good intentions.

In practice, co-rumination amounts to tossing fresh logs onto the fire of an already flaming inner voice. The rehashing of the narrative revives the unpleasantness and keeps us brooding. While we feel more connected and supported by those who engage us this way, it doesn't help us generate a plan or creatively reframe the problem at hand. Instead, it fuels our negative emotions and biological threat response.

Harmful co-ruminative dynamics emerge out of otherwise healthy, supportive relationships because our emotional, inner-voice mechanics aren't actually like a hydraulic system, as Freud and Aristotle and conventional wisdom suggest. Letting out steam doesn't relieve the pressure buildup inside. This is because when it comes to our inner voice, the game of dominoes provides a more appropriate metaphor.

When we focus on a negative aspect of our experience, that

tends to activate a related negative thought, which activates another negative thought, and another, and so on. These dominoes continue to hit one another in a game where there is a potentially infinite supply of tiles. That is because our memories of emotional experiences are governed by principles of *associationism,* which means that related concepts are linked together in our mind.

To illustrate this idea, take a moment to imagine a cat. When you read the word "cat," you probably thought of cats you have known or seen, or actually pictured them in your mind. But you also had thoughts and images of purring sounds, soft fur, and, if you're allergic like me, bouts of sneezing. Now take this associative neural dominoing and apply it to the realm of talking about our emotions. It means that when our friends and loved ones ask us to recount our troubles in detail, related negative thoughts, beliefs, and experiences also spring to mind, which reactivate how bad we feel.

The associative nature of memory, combined with the bias we have to prioritize emotional needs over cognitive ones when we're upset, is why talking often fails to lift our troubled internal dialogues into a more tranquil state. This is one possible explanation for why the Northern Illinois and Virginia Tech students who actively shared their thoughts and feelings about the shootings with other people didn't get any measurable long-term benefit from doing so. And it's why people in the national survey after 9/11 who shared their feelings may have ended up suffering from more physical and mental ills. All of which, of course, raises an urgent question: What's the solution to co-rumination making us feel worse?

Kirk or Spock?

The common shorthand in psychology circles for the tension between emotion and cognition—between what we feel and what we think—is to use the *Star Trek* characters of Captain Kirk and Officer Spock. Kirk is all heart, a man of intense and compelling emotions. He's fire. By contrast, Spock, that lovable, pointy-eared half human half Vulcan, is all head; he's a cerebral problem solver unencumbered by the distractions of feelings. He's ice.

The key to avoid rumination is to combine the two Starship *Enterprise* crew members. When supporting others, we need to offer the comfort of Kirk and the intellect of Spock.

The most effective verbal exchanges are those that integrate both the social and the cognitive needs of the person seeking support. The interlocutor ideally acknowledges the person's feelings and reflections, but then helps her put the situation in perspective. The advantage of such approaches is that you're able to make people who are upset feel validated and connected, yet you can then pivot to providing them with the kind of big-picture advice that you, as someone who is not immersed in their chatter, are uniquely equipped to provide. Indeed, the latter task is critical for helping people harness their inner voice in ways that lead them to experience less chatter over time.

Time, of course, plays a role in our ability to offer perspective-broadening support to the people in our lives. Studies consistently show that people prefer to not cognitively reframe their feelings during the very height of an emotional experience when emotions are worked up; they choose to engage in more intellectual forms of interventions later on. This is where a certain art in talking to other people comes into play, because you must walk a tightrope to take upset people from addressing their emotional needs to the more practical cognitive ones.

As it turns out, one version of this balancing act was codified decades ago by the New York Police Department Hostage Negotiations Team, which emerged in the early 1970s after a series of disastrous situations not just in New York City but also worldwide. To name just a few: the 1971 Attica prison riot, the 1972 Munich Olympics massacre, and the 1972 Brooklyn bank robbery featured in the film *Dog Day Afternoon*. A police officer and clinical psychologist named Harvey Schlossberg was tasked with creating the playbook for the new unit, whose unofficial motto became "Talk to me." Along with prioritizing the need for compassionate engagement over the use of force, he stressed patience. Once the hostage takers understood that they weren't in immediate danger, their autonomic threat response (presumably) eased. This reduced the negative frenzy of their inner voice, allowing the negotiator to shift the dialogue toward ending the standoff.

As soon as the NYPD Hostage Negotiations Team was up and running, the city saw an immediate decrease in bad outcomes for hostage situations. This breakthrough spurred law-enforcement agencies around the globe to follow suit, including the FBI. The bureau developed its own approach called the Behavioral Change Stairway Model, a progression of steps to guide negotiators: Active Listening → Empathy → Rapport → Influence → Behavioral Change. In essence, it's a road map for satisfying people's social-emotional needs that nudges them toward a solution drawing on their cognitive abilities. While law-enforcement negotiators are naturally trying to defuse dangerous situations and arrest criminals, their work bears some similiarities to coaching someone we care about through a problem. In both cases, there is a person who can benefit from the right kind of verbal support.

While all of these strategies apply to how you help the people in your life manage their inner voices, they can also help you make better choices when selecting the people you go to for emotional support. After they've made you feel validated and understood, do they guide you toward brainstorming practical solutions? Or do they excessively extract details and revive the upsetting experience by repeating things like "He's such a jerk! I can't believe he did that." By reflecting after the fact, you can often determine if someone helped you immerse or distance. Most likely, it'll be a combination of the two, which can be a starting point for a dialogue about how the person can better help you next time. By thinking through other experiences with your "chatter advisers," you can also narrow in on which people are right for which problems.

While some friends, colleagues, and loved ones will be useful for a broad range of emotional adversities, when the problems are more specialized, specific people may be more helpful. Your brother might be the right person to coach you through family drama (or, perhaps just as likely, he might be the wrong person). Your spouse might be the perfect chatter adviser for professional challenges, or maybe it's that person from another department at work. Indeed, research indicates that people who diversify their sources of support—turning to different relationships for different needs—benefit the most. The most important point here is to think critically after a chatter-provoking event occurs and reflect on who helped you—or didn't. This is how you build your chatter board of advisers, and in the internet age we can find unprecedented new resources online.

A powerful example is the case of the journalist, sex columnist, and activist Dan Savage and his partner, Terry Miller, who in September 2010 were looking for a way to respond to the news of yet another gay teenager committing suicide after relentless bullying. This time it was a fifteen-year-old named Billy Lucas; he had hanged himself in his grandmother's barn in Greensburg, Indiana. Savage had blogged about his death, and a reader had left a comment saying that he wished he could have told the boy that things—his life—would get better. This prompted Savage and Miller to film themselves talking about how, though their teenage years were hard, they lived happy lives as adults, filled with love and a sense of belonging. They posted the video, and within a week it went viral. Thousands of people made similar videos, and gay teens across the country wrote to Savage to say how it was making them feel more hopeful.

Ten years later—as of this writing—the sentiment that drove that first video is much more than a mere viral phenomenon. It Gets Better is an innovative nonprofit organization and global grassroots movement. More than seventy thousand people have shared their inspiring stories, almost ten times more have pledged support, and untold numbers of young gay people have found comfort, strength, and reasons not to end their lives before they've truly begun. It Gets Better has rescued the inner voices of so many emotionally vulnerable people because, in essence, it acts as a distancing tool promoting normalization—everybody gets picked on, but we all get through it—and mental time travel. Most fascinating of all is the fact that people who watch the video don't have to actually know the speakers to benefit from their advice, a principle that applies to all sorts of similar social-support videos available online. We can find people to coach us through our chatter in the form of prerecorded strangers.

Our discussion of whom we go to for support and how they verbally engage us when we're dealing with chatter raises a question about therapy and its effectiveness, because it obviously involves lots of talking. Does the talking cure, as it is sometimes called, truly cure?

The first thing to keep in mind is that there are countless forms of talk therapy and they often differ drastically in approach. Many empirically validated forms of therapy such as cognitive behavioral therapy employ precisely the kinds of techniques we've been talking about throughout this chapter; they provide clients with emotional support while also crucially helping them engage in cognitive problem solving.

Yet some interventions continue to focus on in-depth emotional venting as a tool for mitigating chatter. Case in point: psychological debriefing, an approach that emphasizes the value of emotional unburdening in the immediate aftermath of negative experiences despite overwhelming evidence arguing against its benefits. The take-home point is that if you find yourself needing more than a conversation with a friend or loved one to deal with your chatter, given what you now know, have a conversation with your prospective mental-health providers to learn about their approach and find out whether it is empirically supported.

Invisible Support

Everything we've explored thus far concerns situations in which people seek support. Yet we all know people who experience chatter and sometimes don't reach out for help. They may be trying to manage a problem on their own or may be concerned about how asking for help might impact the way others view them, or how they see themselves. But often we still want to provide support in some way. After all, observing those we care about in need is a powerful neurobiological experience. It triggers empathy, which motivates us to want to act on their behalf.

Under such circumstances, however, caution is needed. Research shows that there's a danger in trying to dole out unsolicited advice, no matter how skilled you are at blending the strengths of Kirk and Spock. When we give advice at the wrong time, this too can backfire.

Think about the archetypal experience of a parent advising a child how to do a math problem she is struggling with. The parent earnestly looks over the problem, sure that a patient, clear explanation is exactly what their kid needs to succeed at the assignment and feel better about herself. It's a cognitive solution that should lead to positive emotion, right? Except it doesn't play out that way. As the parent explains, the child turns surly and gets agitated. The clean mathematical logic somehow gets lost in emotional static as an argument breaks out.

"I know how to do it!" the kid says.

"But you were having trouble, so that's why I was trying to help," the parent responds.

"I don't need your help!"

The kid storms off to her room. The parent is baffled. What just happened?

(Note: This might or might not have been an autobiographical experience.)

Offering advice without considering the person's needs can undermine a person's sense of *self-efficacy*—the crucial belief that we are capable of managing challenges. In other words, when we are aware that others are helping us but we haven't invited their assistance, we interpret this to mean that we must be helpless or ineffective in some way—a feeling that our inner voice may latch on to. A long history of psychological research into self-efficacy has shown that when it is compromised, it damages not only our self-esteem but also our health, decision making, and relationships.

In the late 1990s, the Columbia psychologist Niall Bolger and his colleagues took advantage of the New York bar exam to examine when people's attempts to provide support for another are most effective. The bar, as all lawyers and their loved ones know, is a grueling, chatter-churning test. Bolger recruited couples in which one person was studying for the bar and, for a little over a month, asked the examinees to answer a set of questions capturing how anxious and depressed they felt, as well as how much support they received from their partner. He also asked the partners of the examinees to report how much support they provided. Bolger was primarily interested in whether benefits that people derive from receiving social support depend on whether a person is aware of the fact that a partner is trying to help.

The study revealed that helping without the recipient being aware of it, a phenomenon called "invisible support," was the formula for supporting others while not making them feel bad about lacking the resources to cope on their own. As a result of receiving indirect assistance, the participants felt less depressed. In practice, this could be any form of surreptitious practical support, like taking care of housework without being asked or creating more quiet space for the person to work. Or it can involve skillfully providing people with perspective-broadening advice without their realizing that it is explicitly directed to them. For example, asking someone else for input that has implications for your friend or loved one in the presence of the person who needs it (a kind of invisible advice) or normalizing the experience by talking about how other people have dealt with similar experiences. Doing these things transmits needed information and support, but without shining a spotlight on the vulnerable person's seeming shortcomings.

Since Bolger's first experiment pioneered this domain, other research has converged to validate the effectiveness of invisible support. A study on marriages, for example, found that partners felt more satisfaction about their relationships the day after receiving invisible support. Another experiment found that people were more successful in meeting their self-improvement goals if the support they received from their partner toward those goals was delivered under the radar.

Further research has yielded insights into the circumstances in which such invisible support is most effective: when people are under evaluation or preparing to be. For example, when they're studying for exams, preparing for interviews, or rehearsing the talking points of a presentation. During such times people feel most vulnerable. In contrast, when people want to manage their chatter as quickly and efficiently as possible, it's not necessary to be subtle or crafty in how you support. In this case, direct advice that blends Kirk and Spock is most needed, appropriate, and likely to succeed.

Along with the forms of invisible support we've discussed, there is one other pathway for subtly aiding people we are very close to who find themselves submerged in chatter, and it's completely nonverbal: affectionate touch.

Touch is actually one of the most basic tools that we use to help those we care most about turn a negative internal dialogue around. Like language, it is inseparable from our ability to manage our emotions from infancy onward, because our caregivers use affectionate physical contact to calm us from the moment we leave the womb. Research shows that when people feel the welcome, affectionate touch or embrace of those they are close to, they often interpret that as a sign that they are safe, loved, and supported. Caring physical contact from people we know and trust lowers our biological threat response, improves our ability to deal with stress, promotes relationship satisfaction, and reduces feelings of loneliness. It also activates the brain's reward circuitry and triggers the release of stress-relieving neurochemicals such as oxytocin and endorphins.

Affectionate touch is so potent, in fact, that one set of studies found that a mere one second of contact on the shoulder led people with low self-esteem to be less anxious about death and feel more connected with others. More striking still, even the touch of just a comforting inanimate object, like a teddy bear, can be beneficial. This is most likely a result of the brain coding contact with a stuffed animal similar to how it codes interpersonal touch. Indeed, many scientists consider the skin a social organ. In this sense, our contact with others is part of an ongoing nonverbal conversation that can benefit our emotions.

What we give to and receive from other people in our daily interactions constitutes a rich portfolio of comfort for the inner voice. The science of how these techniques work is becoming increasingly clear, though of course employing them with people we love takes a certain art, not to mention practice.

Ultimately, the conversations we have with others aren't all that different from the conversations we have with ourselves. They can make us feel better or worse. Depending on how we engage other people, and how they engage us, we experience more or less chatter. This has likely been the case since our species started sharing its problems. We just didn't understand the underlying psychological mechanisms until recently.

Yet in our young twenty-first century, our relationships have begun to migrate to a novel environment for our species and our chatter, the same place the students at Northern Illinois and Virginia Tech went in the wake of their respective tragedies: the internet. A natural question is if the ways in which verbal support succeeds and fails carry over to how we "talk" on social media, over texts, and through other forms of digital communication.

While psychology is only just beginning to grapple with this question, we're already seeing some clues. For example, in the mid-2010s, my colleagues and I wanted to better understand the nature of co-rumination via social media, so we asked people who were in the midst of grappling with an upsetting experience to chat with another person via a computer messaging app. What they didn't know was that the other person was an actor who had been carefully trained to nudge half of the participants to keep talking about what happened. For the other half, he gently encouraged them to zoom out and focus on the bigger picture.

Sure enough, the participants who were led to rehearse their feelings became increasingly upset during the conversations. Their negative emotions skyrocketed from the time they sat down at the keyboard until they left. In contrast, the participants whom the actor helped to zoom out remained just as calm and collected as they were when they first came into the lab.

The thing we don't often think about when we seek or give support, online or off-, is that, objectively speaking, the people in our lives form a *social environment*. What we've been learning

is how to navigate that environment to maximize positive outcomes for the inner voice. Our surroundings are inseparable
from the human beings who inhabit them, and when we use
the resources that are available to us in our relationships with
others, the benefits are powerful. But other people are only one
facet of our environment that we can harness to improve our
internal conversations.

We can also go outside for a walk, attend a concert, or simply tidy up our living space, and each of these seemingly small
actions can have surprising effects on our chatter.

CHAPTER SIX

Outside In

*I*n 1963, the Chicago Housing Authority completed construction on a monumental project on the city's historically black South Side: the Robert Taylor Homes. A vast syndicate of twenty-eight sixteen-story concrete towers, it was the biggest public housing complex in the history of the world.

Built to halt the rise of slum conditions that were taking over more and more neighborhoods, the Robert Taylor Homes were named after a prominent, recently deceased black community leader and architect. Unfortunately, the final product didn't honor his memory. Not only did the Robert Taylor Homes reinforce the citywide structure of segregation that already reigned in Chicago, they gradually exacerbated the challenges facing the community.

By the 1980s, the Robert Taylor Homes had become notorious as a microcosm of the same problems plaguing dozens of American cities: gang violence, drugs, and people beset with fear, ill health, and disenfranchisement. A grand, much-touted

attempt at urban renewal had crumbled into yet another example of urban decline that disproportionately affected African Americans.

If you lived in the Robert Taylor Homes, you didn't have to
turn on the television or read a newspaper to witness the devastating effects that poverty and segregation were having on
America during the second half of the twentieth century. You
simply had to walk outside your apartment. But within this
atmosphere of crime, amid the daily tumult that defined the
lives of the Robert Taylor Homes residents, a groundbreaking
experiment would soon take place.

When people applied for an apartment in the Robert Taylor Homes, they had no say over the building where they
would live. They were randomly assigned to a unit in almost
the same way that scientists randomly assign subjects to different groups in an experiment. As a consequence, tenants found
themselves living in apartments that, in many cases, looked out
onto dramatically different landscapes. Some units faced court-
yards filled with grass and trees. Others looked out onto gray
slabs of cement.

In the late 1990s, this unique circumstance ended up providing Ming Kuo, a newly minted assistant professor working
at the University of Illinois, with an unexpected opportunity.
With short dark hair, glasses, a warm smile, and a penetrating
mind, Ming was interested in understanding whether the physical surroundings of residents affected their ability to cope with
the stress of living in a drug- and crime-filled environment.
Like many other scientists, she had been struck by a growing
body of research that demonstrated a link between views of
green spaces and increased resiliency.

In one particularly compelling study, the environmental
psychologist Roger Ulrich had found that patients recovering

from gallbladder surgery who were assigned to a room that faced onto a small cluster of deciduous trees recovered faster from their operations, took fewer painkillers, and were judged as more emotionally resilient by the nurses caring for them than patients whose rooms looked out onto a brick wall. But whether glancing at green views would help people manage the emotional turmoil of inner-city life in one of the most hostile environments in the United States was a complete mystery.

When Ming learned about the housing assignment process at the Robert Taylor Homes, she saw a chance to further examine the effects of nature on the mind. So she and her team got to work visiting apartments to see what they could uncover. First, they took pictures of the areas surrounding eighteen Robert Taylor Homes buildings and coded each building's view for the presence of green space. Then they went door-to-door recruiting participants for their study; in this case, female heads of households. During forty-five-minute sessions held in the participants' apartments, Ming's team cataloged how well they were managing the most important issues in their lives: whether to go back to school, how to keep their homes safe, and how to raise their children. They also measured each person's ability to focus their attention by measuring how many digits they could retain and manipulate from a string of numbers.

When Ming and her team analyzed the data, they found that the tenants who lived in apartments with green views were significantly better at focusing their attention than those whose buildings looked out onto barren cityscapes. They also procrastinated less when making challenging decisions and felt that the obstacles they faced were less debilitating. In other words, their behavior was more positive; their thinking was calmer and more challenge oriented. What's more, Ming's findings suggested that the Robert Taylor Homes residents' behavior and thinking were more positive *because* they were better able to focus their attention. Trees and grass seemed to act like mental vitamins that fueled their ability to manage the stressors they faced.

As it turned out, Ming's findings were not a fluke. In the years since her study, more green revelations have followed. For example, using data from more than ten thousand individuals in England collected over eighteen years, scientists found that people reported experiencing lower levels of distress and higher well-being when living in urban areas with more green space. Meanwhile, a 2015 high-resolution satellite imagery study of the Canadian city of Toronto found that having just ten more trees on a city block was associated with improvements in people's health comparable to an increase in their annual income of $10,000 or being seven years younger. Finally, a study involving the entire population of England below the age of retirement—approximately forty-one million people—revealed that exposure to green spaces buffered people against several of the harmful effects of poverty on health. To put it another way that only slightly exaggerates, green spaces seem to function like a great therapist, anti-aging elixir, and immune-system booster all in one.

These findings raise a fascinating possibility: that the internal conversations we have with ourselves are influenced by the physical spaces we navigate in our daily lives. And if we make smart choices about how we relate to our surroundings, they can help us control our inner voice. But in order to understand how this works, we first need to know which facets of nature appeal to us.

The Force of Nature

In a certain sense, Ming's work in Chicago with the Robert Taylor Homes didn't start with her or Ulrich's work on gall-bladder patients. Rather, it emerged out of a scientific husband-and-wife duo's curiosity about the interaction between the human mind and the natural world.

In the 1970s, Stephen and Rachel Kaplan, both psychologists at the University of Michigan, had begun to advance an intriguing idea: that nature could act like a battery of sorts, recharging the limited attentional reserves that the human brain possesses. They called it attention restoration theory.

Sure, most people knew that a painterly sunset, mountain view, walk in the woods, or day at the beach usually left a person feeling good, but was there more to it? The Kaplans thought that there was because of a distinction related to human attention that William James, one of the founders of modern-day psychology in the United States, put forth more than a hundred years ago. James separated the ways we paid attention into two categories: involuntary and voluntary.

When we involuntarily pay attention to something, it's because the object of our attention has an inherently intriguing quality that effortlessly draws us to it. In a real-life scenario, you can imagine, say, a talented musician playing on a street corner while you're walking around a city, and you feel yourself notice the sound and gravitate toward it to stop and listen for a few minutes (and then maybe toss some money into the instrument case before you keep walking). Your attention has been gently reeled in by a process the Kaplans called "soft fascination."

Voluntary attention, in contrast, is all about our will. It captures the amazing capacity that we human beings have to shine our attentional spotlight on whatever we want—like a difficult math problem or dilemma we're trying to stop ruminating on. As a result, voluntary attention is easily exhausted and needs continual recharging, while involuntary attention doesn't burn as much of our brain's limited resources.

The Kaplans believed that nature draws our involuntary attention because it is rife with soft fascinations: subtly stimulating properties that our mind is pulled to unconsciously. The natural world delicately captures our attention with artifacts such as big trees, intricate plants, and small animals. We may glance at these things, and approach them for greater appreciation like that musician playing on the corner, but we're not carefully focusing on them as we would if we were memorizing talking points for a speech or driving in city traffic. Activities like those drain our executive-function batteries, whereas effortlessly absorbing nature does the opposite: It allows the neural resources that guide our voluntary attention to recharge.

The studies that Ming and her colleagues went on to perform in Chicago were designed to rigorously test the Kaplans' ideas, and as we've already seen, they produced dramatic supporting evidence. Other experiments have likewise illustrated nature's powers.

One now classic study was done in 2007 just a few blocks from my home in Ann Arbor, when Marc Berman and his colleagues brought participants into the lab and had them perform a demanding test that taxed their attentional abilities—they heard several sequences of numbers that varied from three to nine digits in length, which they were asked to repeat in *backward* order. Half of the participants then went out for a walk in the local arboretum for just under one hour, while the other half walked down a congested street in downtown Ann Arbor for the same amount of time. Then they came back to the lab and repeated the attention task. A week later they swapped circumstances; each person had to go on the walk not taken the previous week.

The finding: Participants' performance on the attention test improved considerably after the nature walk but not the urban walk. Their ability to invert strings of numbers and repeat them back to the experimenter was much sharper. Moreover, the result didn't depend on whether participants took their walks during the idyllic summer or gloomy winter. No matter what time of year, the nature stroll helped their attention more than the urban one did.

Berman and his colleagues went on to replicate these results in other populations. For example, one study with clinically depressed participants indicated that the nature walk improved their cognitive function and led them to feel happier. Another satellite-imagery study conducted by a different team with more than 900,000 participants found that children who grew up with the least exposure to green spaces had up to 15 to 55 percent higher risk of developing psychological disorders such as depression and anxiety as adults. All of this, along with Ming's work in Chicago, suggested that the benefits of nature weren't limited to our attentional reserves. They also extended to our emotions.

Nature's impact on human feelings made sense given how critical the ability to maintain our attention is for helping people manage their inner voice. After all, many of the distancing techniques we've examined rely on focusing the mind; it's hard to keep a journal, "time travel," or adopt a fly-on-the-wall perspective if you can't concentrate. Moreover, the ability to divert our internal conversations away from things that are bothering us, or reframe how we're thinking about stressful situations, likewise requires that our executive functions not be running on empty. But Ming and other scientists never tested the idea that nature could reduce rumination directly. This occurred in 2015 at Stanford University in Palo Alto, California.

Leafy, suburban Palo Alto is a far cry from gritty, crowded Chicago, though it does have a handful of busy streets. Researchers there designed an experiment that had participants take a ninety-minute walk either on a congested avenue or through a green space adjacent to Stanford's campus. When the scientists compared people's rumination levels at the end of the study, they found that participants in the nature-walk group reported experiencing less chatter and less activity in a network of brain regions that support rumination.

As a born and bred city dweller, I think it's necessary to pause here for a moment. Over the last two centuries, human civilization has seen a vast migration from rural areas to urban ones, and by 2050 an estimated 68 percent of the world's population is expected to live in cities. If you have a citified life, it's natural to feel alarmed if you're part of this huge swath of humanity with less access to nature and green spaces. When I first learned about this research, I was certainly disconcerted. I wondered, does having lived in the dense, concrete cities of Philadelphia and New York for the first twenty-eight years of my life mean that I—as well as everyone else with similar urban living experiences—am destined to have worse health, impaired attention, and more ruminative thoughts?

Thankfully, the answer is no. You don't need to be surrounded by nature to "green" your mind. Recall that the underlying idea in the Kaplans' attention restoration theory is that the subtle perceptual features of nature act as a battery of sorts for the brain. Well, the visual characteristics that create this pleasing soft fascination don't just have this effect when you're physically close to nature. Secondhand exposure to the natural world through photos and videos also restores attentional resources. This means that you can bring nature and its sundry benefits into your urban environment—or any environment, for that matter—by glancing at photos or videos of natural scenes. Virtual nature is, incredibly, still nature as far as the human mind is concerned.

An experiment published in 2016, for example, induced stress in participants using the dreaded speech task. Afterward, they watched a six-minute video of neighborhood streets that varied in their green views. At the low end, participants watched a video of homes on a street without any trees; at the high end, the video toured a neighborhood with a lush canopy of trees. Those exposed to the most views of nature demonstrated a 60 percent increase in their ability to recover from the stress of the speech compared with those who saw videos with the fewest views of green spaces.

While the bulk of the research done on the psychological benefits of nature focuses on visual exposure, there's no reason to think that our other senses don't also provide pathways for these startling effects. In 2019, a study found that exposing people to natural sounds, such as rainfall and crickets chirping, improved performance on an attentional task. Such sonic forms of nature may also constitute a soft fascination.

Collectively, these findings demonstrate that nature provides humans with a tool for caring for our inner voice from the outside in, and the longer we're exposed to nature, the more our health improves. It offers us a playbook for structuring our environments to reduce chatter. And bringing new technologies to bear will likely make it easier to reap the benefits. For instance, Marc Berman and his collaborator Kathryn Schertz have developed an app called ReTUNE, short for Restoring Through Urban Nature Experience. It integrates information concerning the greenness, noisiness, and crime frequency of every city block in the neighborhood surrounding the University of Chicago to come up with a naturalness score. When users input their travel destination, the app generates directions that maximize the restorative nature of the walk, taking into account such practical issues as number of road crossings and length of the walk. If proven effective, a natural next step would be to extend the app to, well, everywhere. Of course, you don't need the app to maximize exposure to nature in your daily life. Just make a careful assessment of the different environments you move through and modify your routes accordingly.

As our mind's relationship with nature demonstrates, the physical world is capable of influencing psychological processes deep within us. But nature's many sources of soft fascination are only one pathway through which we reap these benefits. There is another feature that helps us control our inner voice, except this tool isn't limited to our surroundings in the natural world. We can also find it at concerts, in museums, and even watching a baby take its first steps.

Shrinking the Self

The excitement that Suzanne Bott felt as she grabbed her paddle and climbed into the raft made her body tingle. For the next four days, she would be paddling down Utah's shimmering Green River with three other rafts of people. During the day they would take in the tawny, castle-like canyon walls. At night they would talk about the day's adventures around a flickering campfire.

Despite first appearances from a cursory glance at the group, this wasn't your average collection of wilderness enthusiasts. Most of the paddlers were military veterans who had seen combat, along with several former firefighters who had been first responders on 9/11. Each person had replied to an advertisement recruiting veterans for an expenses-paid journey on the Green River designed to help them connect with nature. There was, however, a catch: The trip doubled as a research experiment. Even so, all the participants had to do was paddle and fill out a few questionnaires.

Bott was the outlier of the group. She wasn't a combat veteran and had no experience dousing fires. In 2000, after spending six years earning her PhD in natural resource management from Colorado State University, she felt burned out by the publish-or-perish culture of academia. So she began working in redevelopment, helping revitalize small towns. But Bott remained mindful all the while of the privileged life she led compared with so many other Americans, including her brother, a senior intelligence officer serving in Iraq. While some people's rumination comes from the things that they do, hers came from what she wasn't doing. She needed a change.

After working stateside for several years, Bott found a job with a State Department contractor in Iraq that supported the new government's efforts to take firmer control of different regions of the country. She landed in Baghdad in January 2007 and spent a year deployed in Ramadi, the Iraqi city *Time* magazine had dubbed "the most dangerous place in Iraq" only a month before she arrived. She spent much of her time there developing a long-term transition strategy for the new Iraqi government, working closely with a small corps of marines and army engineers. Her commute involved donning body armor, traveling in Humvee convoys, and sprinting from vehicles to buildings to avoid sniper fire. She was a world away from cozy Colorado.

Her new career provided Bott with the sense of purpose that had been missing from her life. It also pushed her to an emotional breaking point. She attended memorials for fallen colleagues on a regular basis and witnessed horrors amid her work that she wasn't prepared for—car bombs, territorial warfare, and assassinations. Carnage became the stuff of everyday life.

In 2010, Bott returned home to the United States, where her chatter took over. Questions about why she survived when so many of her colleagues had not were a continual source of distress. Memories of the horrors she witnessed replayed in her mind, compounded by constant news reports detailing the rise of ISIS in areas where she had so recently lived and worked. Her chatter reached a crescendo in 2014 when she learned that in Syria ISIS had decapitated James Foley, a journalist she worked closely with in Iraq. Against her own better judgment, she watched the decapitation video that ISIS posted on the internet. She hadn't been the same since. Then she saw the advertisement for the rafting trip.

During the evening after their first day out on the water, Bott filled out a brief questionnaire that asked her to rate how much she experienced several different positive emotions. A team of scientists, led by a psychologist named Craig Anderson from the University of California, Berkeley (who was also participating in the trip), were hoping to use the paddlers' responses to understand the impact of the common but grossly understudied emotional experience of awe.

Awe is the wonder we feel when we encounter something powerful that we can't easily explain. We are often flooded by it in the natural world when we see an incredible sunset, mile-high mountain peak, or beautiful view. Awe is considered a self-transcendent emotion in that it allows people to think and feel beyond their own needs and wants. This is reflected in what happens in the brain during awe-inspiring experiences: The neural activity associated with self-immersion decreases, similar to how the brain responds when people meditate or take psychedelics like LSD, which are notorious for blurring the line between a person's sense of self and the surrounding world.

The feeling of awe, however, is by no means restricted to nature and the great outdoors. Some people experience it when they see Bruce Springsteen in concert, read an Emily Dickinson poem, or take in the *Mona Lisa* at the Louvre. Others may have awe-drenched experiences when they see something extraordinary in person, like a high-stakes sports event or a legendary object such as the U.S. Constitution, or witness something intimately monumental, like an infant taking its first steps. Evolutionary psychologists theorize that we developed this emotion because it helps unite us with others by reducing our self-interest, which provides us with a survival advantage because groups fare better against threats and can achieve loftier goals by working together.

But the Berkeley team wasn't solely interested in whether shooting down roaring rapids would lead the rafters to experience awe. They figured it would. What they really wanted to know was whether the amount of awe they experienced on the trip would have any lasting impact on their stress and well-being after it was over.

So, at the beginning of the rafting trip and a week after it was complete, Anderson asked the rafters to fill out a set of measures indexing their levels of well-being, stress, and PTSD. Much had happened between the two assessments. They rafted dozens of miles over the course of their four-day trip, spent numerous afternoons hiking along the riverbanks, and looked at prehistoric petroglyphs created thousands of years ago that led them to muse on the forgotten societies that had once trod the same ground beside the river as they did now. Would the effects of these experiences dissipate after the trip, or would they leave something behind?

When Anderson crunched the numbers after the study ended, he found that participants displayed significant improvements on each of the well-being measures after the trip ended; their stress and PTSD levels declined, while their overall levels of happiness, satisfaction with life, and sense of belonging improved. Those were interesting results on their own. But the most fascinating finding concerned what predicted them. As Anderson and his colleagues expected, it wasn't a function of how much amusement or contentment or gratitude or joy or pride that paddlers felt during each day of the rafting trip. It was how awe inspiring it felt. Suzanne felt all of these boosts, including a more tranquil inner voice. "That rafting trip changed my perspective dramatically," she told me two years later.

When you're in the presence of something vast and indescribable, it's hard to maintain the view that you—and the voice in your head—are the center of the world. This changes the synaptic flow of your thoughts in similar ways as other distancing techniques we've examined. In the case of awe, however, you don't have to focus your mind on a visual exercise or on reframing an upsetting experience. In this sense, it's similar to saying your own name: You just have the experience, whatever it happens to be, and relief follows. When you feel smaller in the midst of awe-inspiring sights—a phenomenon described as a "shrinking of the self"—so do your problems.

The Berkeley Green River white-water rafting study is just one example of a burgeoning line of research linking awe to physical and psychological benefits. Another study, for instance, showed that awe leads people to perceive time as being more available, pushing them to prioritize time-intensive but highly rewarding experiences like going to a Broadway show over less time-intensive—but also less rewarding—material ones like purchasing a new watch. Meanwhile, on the physiological level, awe is linked with reduced inflammation.

The influence of awe on behavior is so strong, in fact, that others can't help but notice it. One set of studies found that "awe-prone" people came across as humbler to their friends. They also reported higher humility and had a more balanced view of their strengths and weaknesses—both hallmark features of wisdom—and more accurately credited the role of outside influences on their successes.

There is an important caveat to consider when thinking about the role that awe plays in our emotional lives. While the bulk of research links it with positive outcomes, scientists have shown that a subset of awe-inducing experiences can trigger negative feelings. Let's call these encounters "awful" in the negative sense: the sight of a tornado, a terrorist attack, or believing in a wrathful God. (Research shows that approximately 80 percent of awe-related incidents are uplifting and 20 percent aren't.) These kinds of experiences are considered awe inspiring in the sense that they, like a majestic sunset, are so vast and complex that we can't easily explain them. The difference is that people perceive them as threatening. And it turns out that when you inject a bit of threat into the awe equation, that can, perhaps not surprisingly, turn thoughts into chatter.

The operative power of awe is its ability to make us feel smaller, nudging us to cede control of our inner voice to a greater grandeur. But there is another lever that our physical environments can pull to improve our internal dialogues that is the opposite of giving in to life's wild vastness—a lever that doesn't help us cede control but rather helps us regain it.

The Nadal Principle

In June 2018, the Spanish tennis superstar Rafael Nadal stepped onto the clay courts of the French Open to battle in the tournament's final match in pursuit of his eleventh championship there. That summer day in Paris, with fifteen thousand fans waiting restlessly to watch a world-class match, he and his opponent, the Austrian Dominic Thiem, came out of the locker rooms, ready to compete. Nadal did what he always does before a match. First, he walked across the court to his bench with a single racket in his hand. Then he took off his warm-up jacket as he faced the crowd, bouncing back and forth vigorously on the balls of his feet. And as usual, he placed his tournament ID card on his bench facing up.

Then the match began.

Nadal jumped ahead right away, winning the first set. After each point, he fiddled with his hair and shirt before the next serve, as if arranging them back in place. During breaks in the action, he sipped a power drink and water and then returned both exactly as they had been—in front of his chair to his left, one precisely behind the other, aligned at a diagonal with the court.

Two sets later, Nadal beat Thiem and left the French Open victorious yet again.

Although you might think that competing against world-class athletes and making sure you don't pull a muscle are the most essential parts of professional tennis, that's not true for Nadal, one of the greatest players in history. "What I battle hardest to do in a tennis match," he says, "is to quiet the voices in my head." And his quirky customs on the court, which many of his fans find amusing but strange, provide him with a perfectly reasonable method of doing so.

By always placing his ID faceup, carefully arranging his water bottles so they are perfectly aligned in front of his bench, and making sure that his hair is just right before a serve, Nadal is engaging in a process called compensatory control; he's creating order in his physical environment to provide him with the order he seeks internally. As he puts it, "It's a way of placing myself in a match, ordering my surroundings to match the order I seek in my head."

This tendency to structure elements in our environment as a buffer against chatter goes beyond contexts in which our performance is being evaluated. It extends to any of the spaces that we occupy. As a result, humans infuse order into their external surroundings—and by extension their minds—in a variety of ways. Some are very similar to Rafael Nadal's. This might explain the global influence of Marie Kondo and her 2014 best-selling book, *The Life-Changing Magic of Tidying Up*. Her philosophy of decluttering our homes by only retaining objects that give us joy is a strategy for influencing how we feel by imposing order on the environment.

But how does the ordering of our surroundings influence what's happening inside our minds? To answer this question, it's crucial to understand the pivotal role that *perceptions of control*—the belief that we possess the ability to impact the world in the ways we desire—play in our lives.

The desire to have control over oneself is a strong human drive. Believing that we have the ability to control our fate influences whether we try to achieve goals, how much effort we exert to do so, and how long we persist when we encounter challenges. Given all this, it is not surprising that increasing people's sense of control has been linked to benefits that span the gamut from improved physical health and emotional well-being, to heightened performance at school and work, to more satisfying interpersonal relationships. Conversely, feeling out of control often causes our chatter to spike and propels us to try to regain it. Which is where turning to our physical environments becomes relevant.

In order for you to truly feel in control, you have to believe not only that you are capable of exerting your will to influence outcomes but that the world around you, in turn, is an orderly place where any actions you engage in will have their intended effect. Seeing order in the world is comforting because it makes life easier to navigate and more predictable.

The need for order in the external world is so strong, one study found, that after recalling a chatter-provoking incident and focusing on their lack of control, participants actually saw illusory patterns in images. In lieu of other avenues for simulating order, their minds led them to imagine the patterns. In another experiment, participants who couldn't control the noise levels in their surroundings were asked to choose either a postcard of a water lily with a black border that conveyed the idea of structure or a similar postcard that lacked a border. On average, they preferred the one with the structured border, another visual shorthand for order.

What scientists have discovered, however, is that just like Nadal we can simulate a sense of order in the world—and by extension in our own minds—by organizing our surroundings and making sure that our physical environments conform to a particular, controllable structure.

The fascinating thing about seeking compensation for chaos in one area (that is, our minds) by creating order in another (that is, the physical environment) is that it doesn't even have to have anything to do with the specific issue that is throwing off our inner voice. This is why imposing order on our environments is so useful; it's almost always easy to do. And the value of engaging in this practice is impressive. For instance, one experiment demonstrated that just reading about the world described as an orderly place reduced anxiety. Unsurprisingly, research indicates that people who live in more disadvantaged neighborhoods—such as the Robert Taylor Homes in Chicago, and likely the areas of Iraq where Suzanne Bott worked—experience more depression, in part because of disorder they perceive in their surroundings.

In contemporary culture, many people view overly frequent attempts to order one's environment as a sign of pathology. Consider, for example, a subset of people with obsessive-compulsive disorder who are strongly motivated to arrange things so they are orderly. What this research on compensatory control suggests is that these people may simply be taking the strong desire people have to establish order in their surroundings—in order to gain a sense of control—to an extreme. There is logic to what they do, even if restraint is lacking.

What makes OCD harmful—a psychological *dis*order—is

that the need that people with this condition display for order in the environment is excessive and interferes with their normal daily functioning. As a parallel case, our need for order can also get out of control in our larger social surroundings. Just look at the recent proliferation of conspiracy theories online, in which the chaos and upheaval of events are attributed to the shadowy (and orderly) plan of diabolical forces. In this case, people are grasping for order through a narrative mechanism, but often to the detriment of others (the conspiracies are, after all, usually false and based on an absence of evidence).

What research on our need for order and the benefits of nature and awe makes clear is how closely intertwined our physical environments are with our minds. They're part of the same tapestry. We're embedded in our physical spaces, and different features of these spaces activate psychological forces inside us, which affect how we think and feel. Now we know not only why we are drawn to different features of our environment but also how we can make proactive choices to increase the benefits we derive from them.

In 2007, the last of the Robert Taylor Homes was demolished. The city had long since moved out all of the residents, and the once famous symbol of urban blight, segregation, and social disorder was set to be redeveloped into a new complex of mixed-income homes and retail and community spaces. Such a positive, orderly transformation would likely be awe inspiring to the people who remember the crime and violence the buildings were once home to.

Whether the new iteration will have green space integrated into its design in a way that benefits its residents is yet to be determined, but the legacy the original complex left behind still reverberates through the history of Chicago, and the history of science. It is a lasting example of how our environments play a pivotal role in shaping what we think, feel, and do and the importance of actively taking control of our surroundings for our own benefit.

For all the power of our environments, though, we don't just gain psychological relief from our surroundings and the things that fill them. As we saw with the need to exert control, there are also specific things we do in our environments that can help us harness our inner voices, but imposing order the way Nadal does is only the beginning. The methods at our disposal are often so strange, and their effects so strong, that they almost seem like magic.

CHAPTER SEVEN

Mind Magic

One morning in 1762, a three-year-old named Maria Theresia von Paradis woke up blind.

The daughter of an adviser to the Holy Roman Empress, Maria Theresia grew up in Vienna and, in spite of her loss of sight, lived a relatively charmed life. Born a musical prodigy, she excelled at the clavichord, a small rectangular keyboard, and organ. Her talent combined with her disability earned her the attention and generosity of the empress, who ensured that she received a pension and the best education available. By the time she was a teenager, she was a celebrated musician, playing at the most exclusive salons in Vienna and beyond. Mozart would even write a concerto for her. Yet Maria Theresia's parents didn't give up on the idea of their daughter regaining her vision.

As she grew up, doctors experimented with a variety of treatments, administering everything from leeches to electric shocks to Maria Theresia's eyes, all to no avail. Her vision didn't return. Even worse, the treatments left her with a host of maladies. By the time she was eighteen, she suffered from bouts of vomiting, diarrhea, headaches, and fainting spells.

Enter Franz Anton Mesmer, a mysterious Vienna-trained physician who had become well-connected among the city's elite. He claimed to have pioneered a medical intervention that could cure a broad range of physical and emotional ills by altering the flow of an imperceptible force that coursed through the universe using magnetic principles alone. Mesmer cured people's conditions by channeling this invisible energy with magnets and his hands. He called this technique animal magnetism. It would later be eponymized as "mesmerism."

In 1777, when she was eighteen, Maria Theresia began undergoing treatment with Mesmer. Over the course of several months, he touched her eyes and body with his magnets, telling her about animal magnetism and how it would heal her. She was a believer, as were her parents, and sure enough her sight miraculously returned. Not all at once, but in fits and starts.

At first, she just saw blurry images. But then she started distinguishing between black and white objects. Eventually, her sense of color came back. While her perception of depth and proportions still lagged, she gradually began to make out human faces. Yet instead of filling her with joy after all these years, they frightened her, especially noses. The visual world had become alien to her. But the change was still incredible. She could finally see again.

Briefly.

Maria Theresia's parents had a dramatic falling-out with Mesmer, which eventually caused their treatment sessions to end. Hearsay had it that her parents were worried their daughter would lose her pension if she fully recovered her sight. Another version suggested Mesmer and Maria Theresia had been caught carrying on an illicit affair. In any case, their time together was over, and amid swirling rumors Mesmer left Vienna. And when the medical master of animal magnetism disappeared from her life, so did Maria Theresia's vision, yet again.

Mesmer's story, however, didn't end there.

After leaving Vienna and relocating to Paris, he opened a clinic and once again ingratiated himself with the upper classes. He even treated King Louis XVI's wife, Marie Antoinette, along with one of his brothers. During the following years, the demand for Mesmer's services was so great that to increase his profits, he devised a method to increase the number of patients he could simultaneously treat: He directed many people to stand or sit shoulder to shoulder around a wooden tub filled with water and tiny shreds of iron that he had magnetized. Metal rods jutted out from the tub, and with music playing quietly in the background, patients applied the rods to the part of their body that was bothering them while Mesmer walked around adjusting the flow of magnetic energy between rod and patient.

The effectiveness of Mesmer's treatment differed depending on the patients he saw, in some cases significantly. Some people experienced small tinges of pain in the affected parts of their body; some convulsed as if they were having seizures. Others simply felt cured. But not everyone saw improvements. Some experienced something else: nothing at all.

Eventually, in 1784, King Louis had heard enough about mesmerism. He ordered a royal commission of scientists to investigate Mesmer's techniques, led by none other than Benjamin Franklin, who was living in Paris as a diplomat at the time. From the outset, the commission was skeptical of Mesmer's claims. They didn't doubt that some people benefited from being mesmerized. They just didn't believe that the cause was an invisible magnetic force.

The commission's investigation did little to alter their opinion. In one experiment, for example, a woman who was a passionate believer in mesmerism sat next to a closed door. On the other side of the door a Mesmer-trained physician actively applied magnetic energy. When the woman didn't know he was on the other side, she showed no signs of being mesmerized. The moment the same physician made his presence known, the woman began to jerk and flail wildly, indicating the treatment's success. Many similar demonstrations followed.

After concluding the investigation, Franklin and his commission published a damning critique of Mesmer's methods. They wrote that the only healing power that they had observed was the one residing within the human mind: that people simply expecting to feel a certain way could produce a positive outcome—not "animal magnetism." While Mesmer was indeed peddling a force that didn't exist, more than two hundred years later we now know that he provided the world with a valuable peek into a unique tool for combating chatter, which scientific research has only recently caught up to: the magical-seeming power of what we believe, and the profound implications it has for our minds and bodies.

Mesmer hadn't discovered animal magnetism. He had simply administered a *placebo*.

From Worry Dolls to Nasal Sprays

Ask most people what a placebo is, and they'll likely tell you that it is, basically, nothing.

Placebos are commonly understood to be a substance—a sugar pill, in many cases—that is used in pharmaceutical research to gauge the effectiveness of an actual drug. In reality, though, a placebo can be anything—not only a pill, but also a person, an environment, even a lucky charm. And what makes placebos so intriguing is that they can make us feel better even though they have no active medical ingredients.

We use placebos in research to verify that a new medicine or procedure has a clear medicinal effect over and above the simple power of suggestion alone. Doing so acknowledges that the mind possesses real healing potential, but placebos aren't something generally thought of as substantial in their own right. They have long been understood as a tool serving a greater purpose, with no separate use of their own.

This completely misses the point.

Ben Franklin, of course, didn't miss this point. He understood that the benefits Mesmer gave to his patients were real even if animal magnetism wasn't. Yet his timeless insight about the mind's role in healing was swallowed by the sensational story of Mesmer himself. This lasted until the mid-twentieth century, when scientists began to question the idea that placebos were merely a foil for research—in essence, nothing. We now know they are very much something: a remarkable testament to the psychologically intertwined nature of belief and healing, and a hidden back door for subduing chatter.

Placebos are part of an ancient human tradition of endowing objects or symbols with "magic." The mythical seal of King Solomon consists of two interlaced triangles and was believed, among other things, to ward off harmful demons. Likewise, long before it became synonymous with Nazism, the swastika was considered a symbol of good fortune. And still today in Guatemala, when children are scared, they are given a set of tiny figurines dressed in traditional Mayan garb called worry dolls whose job is to take their concerns away.

Many people also develop their own idiosyncratic lucky charms. For instance, the model Heidi Klum carries a tiny bag filled with her baby teeth when she flies and clutches it during turbulence. (Weird, I know, but it helps her.) Michael Jordan wore his college shorts beneath his Chicago Bulls uniform during every game. Of late the healing practice of crystals has become big business—a billion-dollar business, in fact. In a broad sense, placebos are very common. We would be mistaken to write off people who cherish charmed objects as misguided. Scientifically speaking, it's quite rational.

Study after study demonstrates that simply believing that a placebo—a charmed object, healing human presence (like a shaman or trusted physician), or special environment—is going to make us feel better actually does. For example, fewer stomach cramps for irritable bowel syndrome patients, less frequent headache attacks for migraine sufferers, and improved respiratory symptoms for asthmatics. Though the amount of relief that placebos provide varies notably across diseases and patients—like Mesmer's patients, some people are more naturally sensitive to placebos than others—in some cases it can be substantial.

Placebos are even effective for Parkinson's disease. In one experiment, scientists injected a promising new chemical treatment into the brains of patients with advanced Parkinson's symptoms. The hope was that doing so would stimulate dopamine production, impoverished levels of which are a root cause of the disease. After the surgeries were performed, the scientists monitored the patients' symptoms over the next two years. At first glance, the findings were encouraging. Participants who received the injection experienced a significant decline in their symptoms. But there was one problem. Participants in a "sham surgery" group who also had their brains drilled into but didn't have the injection—a placebo, in this context—experienced the same decline in symptoms. They thought they had received the special treatment, so their brains and bodies responded as if they had. The message from this and many other studies is clear: Our minds are sometimes as powerful as modern medicine.

But what about chatter? After all, Mesmer also cured patients suffering from "hysteria," a term that was once used to describe people who experienced difficulty controlling heightened emotions. The placebo of animal magnetism had helped them too. So, do placebos help the inner voice? This was the question that I began discussing one day over coffee with the neuroscientist Tor Wager in 2006, when I was still in graduate school and he was a newly minted assistant professor at Columbia.

"What if we asked people to inhale a nasal spray filled with saline?" he said. "We'll tell them it's a painkiller. I bet that would make them feel better. And we'll also look at their brains."

I won't say I thought Tor was crazy, but I was skeptical at first. Nonetheless, we soon went ahead with the experiment.

The result was the study in which we brought the heartbroken of New York City into the lab to study their brains. You'll recall that we discovered a fascinating overlap between the experience of emotional and physical pain by monitoring participants' brain activity as they looked at a photo of the person who dumped them. But that was only the first part of the experiment.

After participants completed that phase of the study, an experimenter in a white lab coat slid them out of the brain scanner and escorted them to a room down the hallway. The experimenter closed the door, then presented half of them with a nasal spray, telling them it contained a harmless saline solution that would improve the clarity of the MRI brain images that we hoped to collect during the next phase of the study. Then participants inhaled the spray twice in each nostril and returned to the scanner for a second round of brain imaging. The other group underwent the exact same procedure with one crucial difference. The experimenter told them that their nasal spray contained an opioid analgesic drug that would temporarily blunt their experience of pain. The salty spray was our placebo.

Both groups inhaled the same saline solution. But half of them believed that they consumed a substance that would relieve their pain. Now we measured the effect.

Participants who thought they received a painkiller reported experiencing substantially less distress when they relived their rejection. What's more, their brain data told a similar story; they displayed significantly less activity in their brain's social pain circuitry compared with people who knew they had inhaled a saline solution. We discovered that placebos can directly help people with chatter. A spray with nothing chemically meaningful in it could work like a painkiller for the inner voice. It was both strange and exciting: Our minds can cause emotional distress while simultaneously and covertly reducing that distress.

The findings from our study complemented other work documenting the benefits of placebo for managing a range of conditions in which chatter features prominently, like clinical manifestations of depression and anxiety. And in many cases the benefits aren't fleeting. For instance, one large analysis of eight studies found that the benefits of consuming a placebo for reducing depressive symptoms endured for several months.

The wide-ranging effects of placebos raise the question of why they work so miraculously. As it turns out, the explanation isn't miraculous at all. It relates to a necessity our brain generates every waking second of our lives: expectations.

Great Expectations

On August 3, 2012, the comedian Tig Notaro took the stage at the club Largo in Los Angeles and performed a set that immediately became legendary. Four days earlier she had learned that she had cancer in both breasts, but this was only the climax of a recent string of misfortunes. She had been severely ill with pneumonia and gone through a horrible breakup, and her mother had died in a fall. None of this was funny in the least, but she grabbed the mic and started talking anyway.

"Good evening," Notaro said. "Hello. I have cancer."

The crowd laughed expectantly, anticipating a punch line.

"Hi, how are you? Is everybody having a good time?" she continued. "I have cancer."

Some people laughed. Others gasped. The joke was that it wasn't a joke.

If part of comedy is about going to uncomfortable places, Notaro did just that. And this was very uncomfortable. But genius that she is, Notaro walked the tightrope between laughing and crying, and got lots of laughs. For instance, she riffed about how her online dating life would change now, infusing it with new urgency. "I have cancer," she said. "Serious inquiries only."

Her set continued in this startling, tragic, brave, and hilarious vein for twenty-nine minutes in all, and eventually vaulted Notaro to a new level of fame and success (and thankfully, she beat her cancer). What I find so illustrative about it is the way it highlights the essential role that expectations play in governing how we function.

Notaro knew she could make people laugh, even when talking about one of the most somber and chatter-inducing topics imaginable. All she had to do was say the right words in the right order, with the right tone, and with the right pauses. She knew how to do this because of how well honed her expectations were—her expectations about what she could do and what the result would be. If we extend this idea, we begin to realize we all rely on expectations every second of our lives.

You walk. You move. You speak. Now think for a second about how you are able to engage in these actions. How do you determine where to put your foot when you walk, where to run to catch a ball, or how to project your voice when you're speaking to a large group? We are able to do these things because we are constantly, both consciously and subconsciously, making predictions about what we expect is going to happen next, and our brain prepares to respond accordingly.

The brain is a prediction machine that is constantly trying to help us navigate the world. The more we are able to bring our prior experiences to bear on what is required of us, the better we should be at this. And this isn't just relevant to our behaviors. It generalizes to our internal experiences in our bodies as well, which is where placebos come into play. They are a hack for harnessing the power of expectations to influence our minds and physical health.

When a doctor tells you that you'll feel better, this provides you with information you can use to predict how you're actually going to fare over time, especially if she has fancy medical degrees, wears a white coat, and talks with authority. That's not a joke. Research shows that features that you might think are peripheral—if a physician wears a lab coat or not, whether she has acronyms attached to her name placards, and even whether the pills you take are referred to as "brand-labeled" or generic—subconsciously strengthen our beliefs.

Over the course of our lives, we develop automatic beliefs about how certain objects and people influence our health. Like Pavlov's salivating dogs, we see a pill and we reflexively expect that consuming it will lead us to feel better, often without even knowing what it is or how it works.

This pathway for expectations, and by extension placebos, is preconscious. It's not a product of careful thought but rather an automatic, reflexive response. Perhaps unsurprisingly, studies show that rodents and other animals respond to placebos via this same automatic channel. This type of response is adaptive. It provides us with very good guesses about how to react quickly and effectively across a variety of situations. Yet we also evolved an additional pathway in the brain that guides our responses: our conscious thought.

When my head aches and I take a painkiller, I remind myself that swallowing the pill will make me feel better. This simple awareness provides my brain with something invaluable: It helps silence all the doubts I may have about whether the headache will ever go away. *What if nothing will help?* I say to myself. *This hurts so bad. What can I do?* Taking the pill offers me hope that my discomfort is going to diminish and in doing so shifts my internal conversation. Indeed, research indicates that these conscious appraisals draw on the same default system in the brain where our inner voice makes its home.

In a simpler sense, what's going on is that I have a belief.

This belief shapes my expectations, which in turn makes me feel better. People tell us things that we later tell ourselves, and we also have experiences we extract ideas from, and this process creates an infrastructure of expectation in the brain. What particular beliefs we have depends on the people we know and the things that happen to us. But what's really going on in the brain that allows this placebo "magic" to happen?

Because our beliefs pertain to so many different kinds of emotions, physiological responses, and experiences, there is no single neural pathway that creates the placebo effect. For instance, while believing that you'll feel less pain is linked with lower levels of activation in pain circuitry in the brain and spinal cord, thinking that you're drinking an expensive wine can increase activation in the brain's pleasure circuitry. Believing you're consuming a fatty (versus healthy) milk shake even leads to lower levels of the hunger hormone ghrelin. In effect, once you believe something, your neural machinery brings it to fruition by increasing or decreasing the activation levels of other parts of the brain or body related to the processes you are forming beliefs about.

Clearly, there are limits to the effects of placebos. You can't completely believe your way out of any malady. Many medical interventions do provide additional value over and above placebos, and we now know that placebo effects tend to be stronger for psychological outcomes (like chatter) than physical ones. But these caveats notwithstanding, the power of placebos is both profound and undeniable. In fact, mounting evidence indicates that placebos can act as enhancers, supercharging the benefits of certain medicines and treatments.

The problem, however, is that the back door of placebos is tricky to access. For one to work, we have to be deceived into believing that we're consuming a substance or engaging in a behavior that has actual healing properties. Outside of research, where people who participate in studies are typically informed about the possibility of receiving a placebo, such lying would be unethical. So we're left in a bind: We can't lie to ourselves about the medicine we take, which means that in the case of placebos we have access to a tool we can't take advantage of.

Or can we?

If placebos are fundamentally about changing beliefs, then what if we could identify other ways of altering people's expectations that don't involve lying? Information from trusted sources is a potent persuasive device. If I want to convince you of something that you're skeptical of, facts and science often help. Ted Kaptchuk and his team at Harvard capitalized on this idea in 2010 when they published a study that shattered how the scientific world thought about placebos.

First, they settled on a common illness that had already been shown to respond well to placebos: irritable bowel syndrome. After Kaptchuk and his colleagues brought participants with IBS into the medical center where they were performing their study, they explained to them what placebos are and how and why they work. In theory, just learning about the placebo pill should have changed the participants' expectations, which in turn should have spurred their IBS symptoms to wane. And that's exactly what happened.

Over the course of the twenty-one-day experiment, the participants who were educated about the science behind placebo effects and then informed that they were taking a placebo displayed fewer IBS symptoms and greater relief compared with people who were educated about placebos and didn't receive any pills. Understanding how a placebo could make their IBS better actually did just that.

Intrigued by the strange new possibilities of nondeceptive

placebos, my lab conducted our own experiment to examine whether Kaptchuk's findings would generalize beyond problems of the bowels to problems of the mind. We used a similar method and divided participants into two groups, one of which was informed about the science of placebos. In effect, we told them, "If you think a substance will help you, it will." Then we gave them a placebo—a nasal spray again—and told them again that if they thought it would make them feel better, it would.

Next, we stirred up their negative emotions by showing them aversive pictures, such as scenes of blood and gore (participants had agreed to view such images in advance). Sure enough, the people in the placebo group experienced less distress. They also displayed less emotional activity in the brain within two seconds of viewing a distressing image.

Several labs have extended this line of research to other conditions. For example, nondeceptive placebos have been shown to improve allergy symptoms, lower back pain, attention deficit hyperactivity disorder, and depression. We still need to perform more work to understand how powerful and long lasting nondeceptive placebo effects can be. But these discoveries open up a new set of possibilities for how people can cope with physical and emotional pain, and demonstrate how powerfully beliefs affect our inner voice and health. They also reveal something else important—the role culture plays in passing down chatter-fighting practices.

Many of our beliefs are transmitted by the culture we come from, such as the expectations we have about doctors and lucky charms and all sorts of other superstitious influences in our environments. In this sense, the families, communities, religions, and other forms of culture that shape us also provide us with tools for dealing with chatter. Yet beliefs aren't the only "magical" tool that our cultures pass down to us. They offer another approach too: rituals.

The Magic of Fishing with Sharks

World War I turned out remarkably well for Bronislaw Malinowski.

A Polish-born, thirty-year-old anthropology student at the London School of Economics, he traveled to New Guinea in 1914 to conduct fieldwork on the customs of native tribes. Soon after he arrived, however, World War I broke out. This put Malinowski in a politically awkward situation, because he was technically behind enemy lines. He was a citizen of the Austro-Hungarian Empire, now at war with Britain. Meanwhile, New Guinea was an Australian territory and thus an ally of Britain's. As a result, Malinowski couldn't travel back to England or home to Poland, but the local authorities decided to let him continue his work. So he sat out the war in the remote Southern Hemisphere, where he embarked on a quest to understand culture and the human mind.

Malinowski's most important work grew out of the two years he spent in the Trobriand Islands, an archipelago near New Guinea, living with the tribes there to experience their culture firsthand. With his glasses, high boots, white clothing, and pale balding head, he stood out from the islanders, who were dark-skinned, shirtless, and chewed betel nuts, a stimulant like coffee, that turned their teeth red. Yet Malinowski succeeded at gaining their acceptance and a deep understanding of their traditions, including the "magic" involved in their fishing practices.

When the islanders went out on fishing expeditions in safe,

shallow lagoons, they simply grabbed their fishing spears and nets, hopped into their canoes, and glided off along the island waterways until they found their preferred spots. But when they fished in the shark-infested unpredictable waters that surrounded the island, the Trobrianders behaved differently. Before setting off, they offered food to their ancestors, rubbed herbs on their canoes, and chanted magical spells. Then they offered more magical incantations when they were out on the open sea.

"I kick thee down, O shark," they intoned in their language, Kilivila. "Duck down under water, shark. Die, shark, die away."

Of course, the Trobrianders weren't actually engaging in magic. The elaborate choreography of preparation that they engaged in before going on dangerous fishing trips transcends the particularities of their tribe. They were doing something entirely practical on an emotional level that speaks to the psychology of human beings.

They were engaging in ritual—another tool for mitigating chatter.

When people are filled with grief, religious institutions prescribe mourning rites to engage in, such as ritual bathing, burying the dead, and having funerals or memorial services. When cadets attending the U.S. Military Academy at West Point experience stress before an exam, they are told that dressing up in their uniform and walking across campus to spin the spurs on the back of a bronze statue of a Civil War general named John Sedgwick will improve their exam performance. We see rituals increasingly finding their ways into the business world as well. When Southwest Airlines rebranded in 2014 with a new heart-shaped logo on the sides of its planes, pilots began touching it as they stepped on board, and this spread throughout the company, presumably as a source of comfort when facing the daily, inescapable risks of flight.

These are all examples of culturally transmitted rituals, but you can probably think of several idiosyncratic rituals you have created on your own, or those of others. The Hall of Fame third baseman Wade Boggs fielded precisely 150 ground balls, ran wind sprints at exactly 7:17 P.M. (before a 7:35 start time), and ate chicken before every game. To cite another example, for thirty-three years Steve Jobs would look at himself in the mirror each morning and ask himself if that day was the last day of his life whether he'd be happy with what he was going to do. Idiosyncratic rituals of this sort are by no means restricted to famous people. In one study, the Harvard organizational psychologists Michael Norton and Francesca Gino found that the majority of rituals people performed after experiencing a significant loss, such as the death of a loved one or the end of a romantic relationship, were unique.

Regardless of whether the rituals we engage in are personalized or collective, research indicates that when many people experience chatter, they naturally turn to this seemingly magical form of behavior and it offers relief for the inner voice.

A study performed in Israel during the 2006 Lebanon conflict found that women living in war zones who ritualistically recited psalms saw their anxiety decrease, unlike those who didn't. For Catholics, reciting the rosary is a similar dampener of anxiety. Rituals can also help with meeting goals. One experiment found that engaging in a ritual before meals helped women who struggled to eat healthier consume fewer calories than women who tried to be "mindful" about their eating.

Rituals also positively influence performance in high-pressure situations like math exams and (much more fun but even more chatter inducing) performing karaoke. One memorable experiment had participants sing the band Journey's epic song "Don't Stop Believin'" in front of another person. Those who did a ritual beforehand felt less anxiety, had a lower heart rate, and sang better than the participants who didn't. Lesson learned: Start believing in rituals.

It's important to point out that rituals aren't simply habits or routines. Several features distinguish them from the more prosaic customs that fill our lives.

First, they tend to consist of a rigid sequence of behaviors often performed in the same order. That's different from a habit or routine, in which case the sequence of steps composing those behaviors can be looser or frequently change. Take one of my daily routines as an example. When I wake up every morning, I do three things: I pop a thyroid pill (my gland is just a teeny bit underactive), brush my teeth, and drink a cup of tea. While my physician would prefer I take my medication first (it metabolizes better on an empty stomach), that doesn't always happen. Some days the tea comes first. On others, I'll brush as soon as I wake up. And that's okay. I don't feel compelled to repeat the sequence of behaviors if I don't do them in a particular order, and I know that their order won't have a significant effect on me for good or ill.

Now let's contrast what I do every morning with what the Australian Olympic swimmer Stephanie Rice does before every race. She swings her arms eight times, presses her goggles four times, and touches her cap four times. She always does this. This progression of behaviors is Rice's personal and peculiar invention, as are lots of other personalized rituals. In fact, the specific steps that compose rituals often have no apparent connection to the broader goal they're aimed at bringing about. For example, it's not clear how Rice tapping her goggles and cap four times will help her swim faster. But it has meaning for her, and this connects us to the second feature of rituals.

Rituals are infused with meaning. They are charged with significance because they have a crucial underlying purpose, whether it's putting a small rock on a cemetery headstone to honor the dead, engaging in a rain dance to nourish crops, or taking Communion. Rituals take on a greater meaning in part because they help us transcend our own concerns, connecting us with forces larger than ourselves. They simultaneously serve to broaden our perspective and enhance our sense of connection with other people and forces.

The reason rituals are so effective at helping us manage our inner voices is that they're a chatter-reducing cocktail that influences us through several avenues. For one, they direct our attention away from what's bothering us; the demands they place on working memory to carry out the tasks of the ritual leave little room for anxiety and negative manifestations of the inner voice. This might explain why pregame rituals abound in sports, providing a distraction at the most anxiety-filled moment.

Many rituals also provide us with a sense of order, because we perform behaviors we can control. For example, we can't control what will happen to our children throughout their lives, and we can protect them only to a limited degree, which is a source of chatter for many parents. But when they are born, we can baptize them or perform any other of a variety of birth rituals that provide us with an illusion of control.

Because rituals are infused with meaning, and often connect to purposes or powers that transcend our individual concerns, they also make us feel connected to important values and communities, which fulfill our emotional needs and serve as a hedge against isolation. This symbolic feature of rituals also often furnishes us with awe, which broadens our perspective in ways that minimize how preoccupied we are with our concerns. Of course, rituals also frequently activate the placebo mechanism: If we believe they will aid us, then they do.

One of the most intriguing aspects of rituals is that we often engage in them without even knowing it. An experiment performed in the Czech Republic found, for instance, that inducing college students to experience high levels of anxiety led them to subsequently engage in more ritualized cleaning behaviors. Similar findings are evident among children. In one experiment, six-year-olds who were socially rejected by their peers were more likely to engage in repetitive, ritual-like behaviors than other children in the study who weren't rejected.

I have personal experience with similar ritualism myself. While I was writing this book, when I was stuck staring at my computer screen with writer's block, my inner stream of thoughts filled with doubts about whether I'd ever finish, I found myself going to the kitchen to do the dishes, wipe down the counter, and then organize the papers strewn across my home-office desk (a new set of behaviors that my wife found strange, though not objectionable given my usual predilection for making messes rather than cleaning them up). Only when I began researching this chapter did I realize that this was my ritual for dealing with the despair of the writing process and my looming deadline.

This organic emergence of rituals is seemingly a product of the brain's remarkable ability to monitor whether we are achieving our desired goals—for our purposes, the goal of avoiding an inner voice that turns painfully negative. According to many influential theories, your brain is set up like a thermostat to detect when discrepancies emerge between your current and your desired end states. When a discrepancy is registered, that signals us to act to bring the temperature down. And engaging in rituals is one way that people can do this.

I should stress that we don't have to wait to be subconsciously prompted to engage in rituals when we experience chatter. We can engage in them deliberately as well, as I now do whenever I'm feeling stuck at work (my kitchen and home office have never been so clean). There are multiple ways to do so. One approach involves creating our own rituals to engage in before or after a stress-inducing event or to help us deal with chatter. Experiments show that cuing people to engage in completely arbitrary acts that are nevertheless rigid in structure has benefits. For instance, in the karaoke study in which participants had to sing Journey's "Don't Stop Believin'," they were asked to draw a picture of their feelings, sprinkle salt on it, count to five aloud, then ball up the paper and toss it in the trash. This mere onetime reliance on a ritual improved their performance.

Rituals that we see people engage in in lab settings, however, are stripped of their cultural meaning, which we know has additional benefits because it provides a sense of awe, social connection, and feelings of transcendence. With this in mind, another ready avenue for taking advantage of rituals when facing chatter is to lean on those transmitted by our cultures—our families, workplaces, and the broader social institutions we belong to. You might draw on your religion and go to a service, or even the quirky but meaningful rituals of your family. For instance, I make waffles for my kids every Sunday morning after I get back from working out at the gym. It doesn't matter where rituals come from or how exactly they form; they just help.

The Magic of the Mind

The power of placebos and rituals doesn't reside in supernatural forces (though some people believe it does, and that in no way diminishes the benefits of such practices). Their benefits lie instead in their capacity to activate chatter-fighting tools that we carry inside us.

Considering their potency, it's interesting to note that while many people develop their own personalized rituals and placebos, the cultures we are a part of provide us with a vast assortment of these techniques. Culture is often compared to the invisible air we breathe, and much of what we inhale are the beliefs and practices that shape our minds and behavior. You can even think of culture as a system for delivering tools to help people combat chatter. Yet our scientific understanding of these tools is continually advancing, which raises a question: How do we spread this newfound knowledge and integrate it into our culture as a whole?

I never truly contemplated this question until I was forced to confront it when a student of mine raised her hand in class.

What she asked me changed everything.

Conclusion

*W*hy are we learning about this *now*?"

These words, asked in exasperation, came from a student named Arielle on the final day of a seminar I was teaching. For the past three months, I had spent my Tuesday afternoons with twenty-eight University of Michigan undergraduates in the basement of the Psychology Department discussing what science has taught us about people's ability to control their emotions, including chatter caused by the inner voice. The students' final assignment was to come to class with questions for me. It was their chance to raise any lingering doubts before the end of the course, and in most cases before graduating and moving on to the next phase of their lives. It was the session I looked forward to the most each semester I taught the class. The discussions always sparked interesting ideas, some of which even led to new studies. Little did I know when I entered the classroom on that sunny afternoon that this particular final session would add a new dimension to my work as a scientist.

As soon as class began, Arielle had shot her hand up, urging me to call on her first. I obliged, but I didn't understand what she was asking. "Can you be more specific?" I said.

"We've spent this whole semester learning ways to feel better and be more successful," she said, "but most of us are going to graduate this year. Why didn't anyone teach us about these things earlier, when we could have really benefited from them?"

After you teach a class a few times, you usually know what questions to expect. But this one was new. I felt as if I had just run face-first into a wall I hadn't known was there.

I deflected Arielle's question onto the rest of the class (yep: classic professor technique). Students started raising their hands and offering ideas. But I was barely listening. I was stuck inside my head, fixated on what she had said.

The truth was, I didn't have an answer.

Eventually, class time wound down, I said goodbye to the students, and off they went into their futures. But what Arielle had been getting at lodged itself in my mind like a splinter.

Throughout my career—and throughout that semester too—I have met people desperate to escape their inner voice because of how bad it makes them feel. This is understandable. As we know, chatter can pollute our thoughts and fill us with painful emotions that, over time, damage all that we hold dear—our health, our hopes, and our relationships. If you think of your inner voice as an inner tormentor, then it's natural to fantasize about permanently muting it. But losing your inner voice is, in fact, the last thing you would ever want if your aim is to live a functional life, much less a good one.

While many cultures today celebrate living in the moment,

our species didn't evolve to function this way all the time. Quite the contrary. We developed the ability to keep our inner worlds pulsing with thoughts and memories and imaginings fueled by the inner voice. Thanks to our busy internal conversations, we are able to hold information in our minds, reflect on our decisions, control our emotions, simulate alternative futures, reminisce about the past, keep track of our goals, and continually update the personal narratives that undergird our sense of who we are. This inability to ever fully escape our minds is a main driver of our ingenuity: the things we build, the stories we tell, and the futures we dream.

It is a mistake, however, to value our own inner voice only when it buoys our emotions. Even when the conversations we have with ourselves turn negative, that in and of itself isn't a bad thing. As much as it can hurt, the ability to experience fear, anxiety, anger, and other forms of distress is quite useful in small doses. They mobilize us to respond effectively to changes in our environments. Which is to say, a lot of the time the inner voice is valuable not in spite of the pain it causes us but because of it.

We experience pain for a reason. It warns us of danger, signaling us to take action. This process provides us with a tremendous survival advantage. In fact, each year a small number of people are born with a genetic mutation that makes it impossible for them to feel pain. They usually end up dying young as a result. Because they don't experience, for instance, the discomfort of an infection, the burn of scalding water, or the agony of a broken bone, they don't know the help they are in need of or their extreme vulnerability.

This phenomenon mirrors the indispensability of the harsh side of our inner voice. It can cloud our thoughts with negative emotions, but if we didn't have this critical self-reflective capacity, we'd have a difficult time learning, changing, and improving. As uncomfortable as it is when I make a joke that bombs at a dinner party, I'm grateful that afterward I can replay what went wrong in my mind so I hopefully won't embarrass myself—and my wife—next time.

You wouldn't want to live a life without an inner voice that upsets you some of the time. It would be like braving the sea in a boat with no rudder.

When Jill Bolte Taylor, the neuroanatomist who suffered a debilitating stroke, experienced her verbal stream crawling to a stop, and along with it her chatter, she felt strangely elated but also empty and disconnected. We need the periodic pain of our internal conversations. The challenge isn't to avoid negative states altogether. It's to not let them consume you.

Which brings me back to my student Arielle.

What she meant when she asked her question was this: Why hadn't she learned earlier in her life how to reduce episodes of full-blown chatter? Of course, she, just like all of us, possessed many of the tools she needed to control her inner voice. But until she took my class, she didn't have an explicit guide for how to manage it, and Arielle's question made me wonder whether we were doing enough to share this knowledge.

A few weeks after that class, my older daughter, who was four at the time, came home from school in tears. She told me that a boy in her class was taking her toys, which was making her feel bad. As she recounted what happened and I tried to comfort her, Arielle's question popped back into my head. Here I was, a supposed expert on controlling emotions, and yet my own daughter was struggling. Granted, she was just four, which is when the neural circuitry underlying the ability to control your emotions is still developing. Nonetheless, the thought troubled me.

I wondered about what she and her friends were learning at school and whether they would develop the tools that Arielle felt had been withheld from her until she took my class. And eighteen years later, would my daughter ask a professor the same question Arielle asked me? Or more likely, she'd ask me, which would make me feel even worse.

During the days and months that followed, I reflected on the rich and startling variety of ways to distance, to talk to oneself, to leverage and improve personal relationships, to benefit from our environments, and to use placebos and rituals to harness the ability of the mind to heal itself. These techniques had been hidden in plain sight inside us and around us. And while no specific tool is a panacea, they all have the potential to bring down the temperature of our inner voice when it runs too hot. But these findings didn't seem to be penetrating into the world.

So I got to work and recruited a group of like-minded scientists and educators to translate what we know about the science of managing emotion into a course that could be weaved into middle and high school curricula.

After traveling around the country and meeting with hundreds of educators and scientists, in the fall of 2017 we launched a pilot study. Its aim was to translate research on controlling our emotions—including how to harness our inner voice—into a curriculum, and to evaluate what the implications of teaching students about this information are for their health, performance, and relationships with others. We call it the Toolbox Project.

And thankfully, our efforts are beginning to pay off.

In the pilot study, a culturally and socioeconomically diverse group of some 450 students from a public school in the United States participated in the toolbox course we designed. The results were exciting: Kids in the toolbox-curriculum classes who learned about techniques such as journaling, distanced self-talk, and challenge-oriented reframing actually used them to a significant extent in their daily lives. And this is just the start. Soon we are planning to run a much larger study with close to twelve thousand students.

The metaphor of the toolbox doesn't just describe the curriculum my colleagues and I developed. It also describes what I hope you take away from this book.

Distancing is a tool, whether it's imagining yourself as a fly on the wall, mentally traveling through time, or visualizing yourself and your predicaments as physically smaller in your mind. So is distanced self-talk: You can talk to yourself or about yourself using non-first-person pronouns or your own name, and you can normalize your challenges with the universal "you." We can be an inner-voice tool for the people grappling with chatter in our lives—and they can do the same for us—by avoiding co-rumination and finding a balance between providing caring support and helping others constructively reframe their problems when their emotions cool. We can also help in invisible ways that ease the strains of people under stress who may feel insecure about their capabilities. These anti-chatter approaches apply to the ways we interact in our increasingly immersive digital lives as well, though there are behaviors online that are just as important to avoid: passive instead of active use of social media, and doing things lacking empathy that we wouldn't do off-line.

Another subset of tools comes from the complex world

around us. Mother Nature is a veritable toolshed for our minds, containing pleasant and effective ways of restoring the attentional tools that are so helpful for reducing chatter and bolstering our health. It can fill us with awe, as can plenty of experiences found not on mountaintops but at concerts, in places of worship, and even in special moments in our own homes (just remembering when each of my daughters said "Dada" for the first time rekindles awe in me).

Imposing order on our surroundings likewise can be comforting and allow us to feel better, think more clearly, and perform at higher levels. Then there are our beliefs, whose malleability can work to our advantage. Through the neural apparatus of expectation, sugar pills that we know are just sugar pills can improve our health, as can the exercise of rituals, both those that are culturally ordained and those we create ourselves. The power of the mind to heal itself is, indeed, magical (in the awe-inspiring, not supernatural, sense).

You now know about these different tools, but it's critical that you build your own toolbox. That is your personal puzzle, and it's why subduing chatter can frequently be so challenging, even when we know the research.

Science has shown us so much, but there is still more to learn.

We have only just begun to understand how the various strategies for controlling chatter work together for different people in different situations, or how they work when used interchangeably. Why do some tools work better for us than others? We each need to discover which tools we find most effective.

Managing our inner voice has the potential not only to help us become more clearheaded but to strengthen the relationships we share with our friends and loved ones, help us offer better support to people we care about, build more organizations and companies where people are insulated against burnout, design smarter environments that leverage nature and order, and rethink digital platforms to promote connection and empathy. In short, changing the conversations we have with ourselves has the potential to change our lives.

My interest in introspection came from my dad, and when people hear the story about how he used to encourage me to "go inside" and "ask yourself the question" during my childhood, they often wonder whether I do the same with my children when they feel upset.

The answer to that question is no. I most certainly do not. I'm not my dad. But that doesn't mean that I don't talk to my kids about how they can address their chatter. As a parent who wants his children to be happy, healthy, and successful, and as a scientist who knows how important harnessing the conversations we have with ourselves is for achieving these goals, I can't think of a more important lesson to teach them. I just do it in my own way.

I stick Band-Aids on their elbows when they're upset and tell them that if they think the Band-Aid will make them feel better, it will. I take them for walks in the splendidly green arboretum near our house when they feel sad, and carefully nudge them to focus on the big picture when they tell me about their latest tiff on the playground or in the classroom. And when they are acting impossibly irrational for the silliest reasons, I ask them to tell themselves what they imagine their mom or I would say to them. And I tickle them.

One of the things that has become clear to me while writing this book is just how influential a role my wife and I play in the conversations that our daughters have with themselves. We ourselves are one of their tools, in the sense that we provide them with chatter support when they need it and we create the culture they are immersed in at home. We are shaping their inner voices, just as they increasingly affect our own.

Often the things I tell my daughters to rein in their chatter help. Sometimes, I'll admit, they roll their eyes the way I sometimes did with my dad. But over time I've noticed that both of them have begun to implement many of these practices on their own, cycling back and forth through the different techniques they have at their disposal in their unique style as they discover what works. In this way, I hope I can help my daughters harness the conversations they have with themselves throughout their lives.

I also remind my daughters and myself that while creating a calming distance between our thoughts and our experiences can be useful when chatter strikes, when it comes to joy, doing the opposite—immersing ourselves in life's most cherished moments—helps us savor them.

The human mind is one of evolution's greatest creations, not just because it allowed our species to survive and thrive, but because in spite of the inevitable pain that comes with life, it also endowed us with a voice in our head capable of not only celebrating the best times but also making meaning out of the worst times. It's this voice, not the din of chatter, that we should all listen to.

I haven't been in touch with Arielle since our last class together, so she doesn't know what her question inspired. She will, though, if she ends up reading this book, which has been the other effort that grew out of that final class meeting. This book is another attempt to share the discoveries that science has revealed but that haven't yet rooted themselves in our culture.

In a certain sense, there are too many Arielles to count out in the world—people hungry to learn about their own minds, how they give rise to chatter, and how it can be controlled.

So I wrote this book for them.

And for myself.

And for you.

Because no one should have to pace his house at 3:00 A.M. with a Little League baseball bat.

The Tools

Chatter reviews the different tools that exist for helping people resolve the tension between getting caught in negative thought spirals and thinking clearly and constructively. Many of these techniques involve shifting the way we think to control the conversations we have with ourselves. But a central idea of this book is that strategies for controlling the inner voice exist outside us too, in our personal relationships and physical environments. Scientists have identified how these tools work in isolation. But you must figure out for yourself which combination of these practices works best for you.

To help you in this process, I've summarized the techniques discussed in this book, organizing them into three sections: tools that you can implement on your own, tools that leverage your relationships with other people, and tools that involve your environment. Each section begins with the strategies that you are likely to find easiest to implement when chatter strikes, building up to those that may require a little more time and effort.

Tools You Can Implement on Your Own

The ability to "step back" from the echo chamber of our own minds so we can adopt a broader, calmer, and more objective perspective is an important tool for combating chatter. Many of the techniques reviewed in this section help people do this, although some—like performing rituals and embracing superstitions—work via other pathways.

1. *Use distanced self-talk.* One way to create distance when you're experiencing chatter involves language. When you're trying to work through a difficult experience, use your name and the second-person "you" to refer to yourself. Doing so is linked with less activation in brain networks associated with rumination and leads to improved performance under stress, wiser thinking, and less negative emotion.

2. *Imagine advising a friend.* Another way to think about your experience from a distanced perspective is to imagine what you would say to a friend experiencing the same problem as you. Think about the advice you'd give that person, and then apply it to yourself.

3. *Broaden your perspective.* Chatter involves narrowly focusing on the problems we're experiencing. A natural antidote to this involves broadening our perspective. To do this, think about how the experience you're worrying about compares with other adverse events you (or others)

have endured, how it fits into the broader scheme of your life and the world, and/or how other people you admire would respond to the same situation.

4. *Reframe your experience as a challenge.* A theme of this book is that you possess the ability to change the way you think about your experiences. Chatter is often triggered when we interpret a situation as a threat—something we can't manage. To aid your inner voice, reinterpret the situation as a challenge that you can handle, for example, by reminding yourself of how you've succeeded in similar situations in the past, or by using distanced self-talk.

5. *Reinterpret your body's chatter response.* The bodily symptoms of stress (for example, an upset stomach before, say, a date or presentation) are often themselves stressful (for instance, chatter causes your stomach to grumble, which perpetuates your chatter, which leads your stomach to continue to grumble). When this happens, remind yourself that your bodily response to stress is an adaptive evolutionary reaction that improves performance under high-stress conditions. In other words, tell yourself that your sudden rapid breathing, pounding heartbeat, and sweaty palms are there not to sabotage you but to help you respond to a challenge.

6. *Normalize your experience.* Knowing that you are not alone in your experience can be a potent way of quelling chatter. There's a linguistic tool for helping people do this: Use the word "you" to refer to people in general when you think and talk about negative experiences. Doing so helps people reflect on their experiences from a

healthy distance and makes it clear that what happened is not unique to them but characteristic of human experience in general.

7. *Engage in mental time travel.* Another way to gain distance and broaden your perspective is to think about how you'll feel a month, a year, or even longer from now. Remind yourself that you'll look back on whatever is upsetting you in the future and it'll seem much less upsetting. Doing so highlights the impermanence of your current emotional state.

8. *Change the view.* As you think about a negative experience, visualize the event in your mind from the perspective of a fly on the wall peering down on the scene. Try to understand why your "distant self" is feeling the way it is. Adopting this perspective leads people to focus less on the emotional features of their experience and more on reinterpreting the event in ways that promote insight and closure. You can also gain distance through visual imagery by imagining moving away from the upsetting scene in your mind's eye, like a camera panning out until the scene shrinks to the size of a postage stamp.

9. *Write expressively.* Write about your deepest thoughts and feelings surrounding your negative experience for fifteen to twenty minutes a day for one to three consecutive days. Really let yourself go as you jot down your stream of thoughts; don't worry about grammar or spelling. Focusing on your experience from the perspective of a narrator provides you with distance from the experience, which helps you make sense of what you felt in ways that improve how you feel over time.

10. *Adopt the perspective of a neutral third party.* If you find yourself experiencing chatter over a negative interaction you've had with another person or group of people, assume the perspective of a neutral, third-party observer who is motivated to find the best outcome for all parties involved. Doing so reduces negative emotions, quiets an agitated inner voice, and enhances the quality of the relationships we share with the people we've had negative interactions with, including our romantic partners.

11. *Clutch a lucky charm or embrace a superstition.* Simply believing that an object or superstitious behavior will help relieve your chatter often has precisely that effect by harnessing the brain's power of expectation. Importantly, you don't have to believe in supernatural forces to benefit from these actions. Simply understanding how they harness the power of the brain to heal is sufficient.

12. *Perform a ritual.* Performing a ritual—a fixed sequence of behaviors that is infused with meaning—provides people with a sense of order and control that can be helpful when they're experiencing chatter. Although many of the rituals we engage in (for example, silent prayer, meditation) are passed down to us from our families and cultures, performing rituals that you create can likewise be effective for quieting chatter.

Tools That Involve Other People

When we think about the role that other people in our lives play in helping us manage our inner voice, there are two issues to consider. First, how can we *provide* chatter support for others? And second, how can we *receive* chatter support our-selves?

Tools for Providing *Chatter Support*

1. *Address people's emotional* and *cognitive needs.* When people come to others for help with their chatter, they generally have two needs they're trying to fulfill: They're searching for care and support, on the one hand (emotional needs), and concrete advice about how to move forward and gain closure, on the other (cognitive needs). Addressing *both* of these needs is vital to your ability to calm other people's chatter. Concretely, this involves not only empathically validating what people are going through but also broadening their perspective, providing hope, and normalizing their experience. This can be done in person, or via texting, social media, and other forms of digital communication.

2. *Provide invisible support.* Offering advice about how to reduce chatter can backfire when people don't ask for help; it threatens people's sense of self-efficacy and au-tonomy. But that doesn't mean there aren't still ways of helping others when they experience chatter and don't ask for assistance. In such situations providing support *in-visibly,* without people being aware you're helping them, is useful. There are many ways to do this. One approach involves covertly providing practical support, like clean-ing up the house without being asked. Another involves helping broaden people's perspectives indirectly by, for example, talking in general terms about others who have

dealt with similar experiences (for example, "It's amazing how stressful everyone finds parenthood") or by soliciting advice from someone else but without signaling that the questions are meant to help the person in need. For example, if my colleague was struggling to connect with their graduate student and we found ourselves at a function with other advisers, I might casually ask a group whether they've experienced trouble connecting with their students and, if so, how they managed the situation.

3. *Tell your kids to pretend they're a superhero.* This strategy, popularized in the media as "the Batman effect," is a distancing strategy that is particularly useful for children grappling with intense emotions. Ask them to pretend they're a superhero or cartoon character they admire, and then nudge them to refer to themselves using that character's name when they're confronting a difficult situation. Doing so helps them distance.

4. *Touch affectionately (but respectfully).* Feeling the warm embrace of a person we love, whether that be holding someone's hand or sharing a hug, reminds us at the conscious level that we have supportive people in our lives whom we can lean on—a chatter-relieving psychological reframe. Affectionate touch also unconsciously triggers the release of endorphins and other chemicals in the brain such as oxytocin that reduce stress. Of course, for affectionate touch to be effective it has to be welcome.

5. *Be someone else's placebo.* Other people can powerfully influence our beliefs, including our expectations about how effectively we can deal with chatter and how long it

will last. You can utilize this interpersonal healing pathway by providing the people you're advising with an optimistic outlook that their conditions will improve, which changes their expectations for how their chatter will progress.

Tools for Receiving *Chatter Support*

1. *Build a board of advisers.* Finding the right people to talk to, those who are skilled at satisfying both your emotional and your cognitive needs, is the first step to leveraging the power of others. Depending on the domain in which you're experiencing chatter, different people will be uniquely equipped to do this. While a colleague may be skilled at advising you on work problems, your partner may be better suited to advising you on interpersonal dilemmas. The more people you have to turn to for chatter support in any particular domain, the better. So build a diverse board of chatter advisers, a group of confidants you can turn to for support in the different areas of your life in which you are likely to find your inner voice running amok.

2. *Seek out physical contact.* You don't have to wait for someone to give you affectionate touch or supportive physical contact. Knowing about the benefits they provide, you can seek them out yourself, by asking trusted people in your life for a hug or a simple hand squeeze. Moreover, you need not even touch another human being to reap these benefits. Embracing a comforting inanimate object, like a teddy bear or security blanket, is helpful too.

3. *Look at a photo of a loved one.* Thinking about others who care about us reminds us that there are people we can turn to for support during times of emotional distress. This is why looking at photos of loved ones can soothe our inner voice when we find ourselves consumed with chatter.

4. *Perform a ritual with others.* Although many rituals can be performed alone, there is often added benefit that comes from performing a ritual in the presence of others (for example, communal meditation or prayer, a team's pregame routine, or even just toasting drinks with friends the same way each time by always saying the same words). Doing so additionally provides people with a sense of support and self-transcendence that reduces feelings of loneliness.

5. *Minimize passive social media usage.* Voyeuristically scrolling through the curated news feeds of others on Facebook, Instagram, and other social media platforms can trigger self-defeating, envy-inducing thought spirals. One way to mitigate this outcome is to curb your *passive* social media usage. Use these technologies *actively* instead to connect with others at opportune times.

6. *Use social media to gain support.* Although social media can instigate chatter, it also provides you with an unprecedented opportunity to broaden the size and reach of your chatter-support network. If you use this medium to seek support, however, be cautious about impulsively sharing your negative thoughts. Doing so runs the risk of

sharing things that you may later regret and that may upset others.

Tools That Involve the Environment

1. *Create order in your environment.* When we experience chatter, we often feel as if we are losing control. Our thought spirals control us rather than the other way around. When this happens, you can boost your sense of control by imposing order on your surroundings. Organizing your environment can take many forms. Tidying up your work or home spaces, making a list, and arranging the different objects that surround you are all common examples. Find your own way of organizing your space to help provide you with a sense of mental order.

2. *Increase your exposure to green spaces.* Spending time in green spaces helps replenish the brain's limited attentional reserves, which are useful for combating chatter. Go for a walk in a tree-lined street or park when you're experiencing chatter. If that's not possible, watch a film clip of nature on your computer, stare at a photograph of a green scene, or even listen to a sound machine that conveys natural sounds. You can surround the spaces in which you live and work with greenery to create environments that are a boon to the inner voice.

3. *Seek out awe-inspiring experiences.* Feeling awe allows us to transcend our current concerns in ways that put our problems in perspective. Of course, the experiences that provide people with awe vary. For some it is exposure to a breathtaking vista. For someone else it's the memory of

a child accomplishing an amazing feat. For others it may be staring at a remarkable piece of art. Find what instills a sense of awe within you, and then seek to cultivate that emotion when you find your internal dialogue spiraling. You can also think about creating spaces around you that elicit feelings of awe each time you glance at them.

Acknowledgments

*T*he seed for *Chatter* was planted thirty-seven years ago, when my dad started encouraging me to "go inside." His voice was a constant companion as I wrote this book.

To my students, collaborators, and colleagues (there are too many of you to name). Without you, there would be no *Chatter*. Working with you has been a privilege. I hope this book allows other people to benefit from your wisdom the way I have.

It's hard to imagine how I could've finished this project without my family's support. My wife, Lara, patiently listened to me talk about *Chatter* every day for several years. She read every word and never stopped cheering me on. I shudder to think where the kids would be without her (likely stranded at school, in tattered clothing, hungry, wondering why I forgot to pick them up). I, too, would be lost. I'm confident that my father-in-law, Basil, had no idea what he was getting into when he offered to provide advice whenever I needed it. Suffice it to say, I took him up on the offer. Thank you for your indefatigable love and support. Mom, Irma, Karen, Ian, Lila, and Owen—thank you for putting up with my absences and not judging me (too) harshly for working on vacation. I love you all.

Doug Abrams, my literary agent extraordinaire, isn't just brilliant and savvy and tall. He has a magnificent heart. His drive to make the world a better place is intoxicating. Doug had a clear vision for *Chatter* before I did, and tirelessly worked to bring the project to life. His voice was another welcome companion throughout the project. Aaron Shulman started out as my writing coach and ended up becoming a close friend. He taught me how to write for a broad audience, unlocked the secrets to finding great stories, punched up my prose when it needed a lift, and helped me push the manuscript across the finish line during the final sprint. He was my consummate literary guide. Lara Love provided incisive feedback on every chapter of the book, patiently explained how the publishing industry works, and spent countless hours schmoozing with me. Her warmth and wisdom made writing *Chatter* fun. Tim Duggan, my editor at Penguin Random House, was a dream to work with. Discerning, patient, and empathic, he championed *Chatter* from the moment we began collaborating and never stopped. His perceptive line edits and gentle encouragements to scale back here and go deeper there transformed the manuscript. I'm eternally grateful we had the opportunity to work together. I hope we get to do it again.

Thinking about all the people who contributed to *Chatter* is moving. Joel Rickett, my UK editor, offered multiple rounds of penetrating feedback. His suggestion to have an "aha" on every page was a guiding mantra while I worked on the book, and his encouragement to investigate how chatter manifests in dreams remains one of my favorite sidebars. Will Wolfslau read every chapter and made countless suggestions that improved *Chatter*'s final form. Aubrey Martinson (and Will) deftly nursed the manuscript through the publication process, keeping me updated on progress every step of the way. Molly Stern championed *Chatter* from the moment she saw the proposal. Rachel Klayman, Emma Berry, and Gillian Blake provided exceptional input on several chapters. Their advice enhanced *Chatter*'s depth and breadth in ways that I'm grateful. Finally, Evan Nesterak is a fact-checking wunderkind. His meticulousness helped me sleep well knowing that every story detail I presented was confirmed.

Idea Architects is a literary agency filled with sharp minds who are passionate about what they do. Thank you, Rachel Neuman, Ty Love, Cody Love, Janelle Julian, Boo Prince, Mariah Sanford, Katherine Vaz, Kelsey Sheronas, Esme Schwall Weigand, and the rest of the team for all your help. At Penguin Random House, Steve Messina, Ingrid Sterner, Robert Siek, Linnea Knollmueller, Sally Franklin, Elizabeth Rendfleisch, Chris Brand, Julie Cepler, Dyana Messina, and Rachel Aldrich. At Ebury, Penguin Random House UK, Leah Feltham and Serena Nazareth. Abner Stein and the Marsh Agency helped spread word about *Chatter* around the world. I'm indebted to the hard work that Caspian Dennis, Sandy Violette, Felicity Amor, Sarah McFadden, Saliann St. Clair, Camilla Ferrier, Jemma McDonagh, and Monica Calignano put into the project along with the rest of both agencies' teams.

Walter Mischel passed away before he could read *Chatter*. His influence permeates its pages. Özlem Ayduk and I have been close friends and research partners since the first day of graduate school. Her everlasting friendship and support motivated me throughout the project. *Chatter* is filled with her wisdom as well.

Angela Duckworth is the busiest scientist I know. Yet, she always returned my calls (usually minutes after I rang) and never failed to provide wise advice and heartfelt encouragement. David Mayer patiently listened to me pitch countless stories on our weekly runs. Jason Moser was a consummate brainstorming partner who provided a keen clinical perspective on several issues I grappled with (in the book, not personally). Little did I know when I met Jamil Zaki in graduate school that we'd end up writing books at the same time. He is the quintessential *Chatter* Adviser.

Adam Grant, Susan Cain, Dan Pink, Dan Heath, Jane McGonigal, Maria Konnikova, Adam Alter, Elissa Epel, Sonja Lyubomirsky, Dave Evans, Tom Boyce, James Doty, John Bargh, Scott Sonenshein, and Andy Molinsky were all tremendous supporters of this project from its inception. Thank you all for your kind words.

Dozens of people generously shared their amazing stories with me. Thank you. Without them, *Chatter* would not be what it is.

I'm fortunate to work with colleagues who are as generous with their time as they are brilliant. John Jonides, Susan Gelman, Oscar Ybarra, Luke Hyde, Jacinta Beeher, Gal Sheppes, Daniel Willingham, David Dunning, Steve Cole, Ariana Orvell, Marc Berman, Rudy Mendoza Denton, Andrew Irving, Ming Kuo, Amie Gordon, Marc Seery, Scott Paige, Lou Penner, Nick Hoffman, Dick Nisbett, Shinobu Kitayama, Stephanie Carlson, Rachel White, Craig Anderson, Janet Kim, Bernard Rimé, Walter Sowden, Philippe Verduyn, and Tor Wager all provided helpful feedback throughout the writing process. I'd also like to acknowledge the University of Michigan, a unique institution that encourages its faculty to ask "big" questions that matter. Without its support, much of the research I talk about in *Chatter* would not have been possible. I am also grateful to the National Institutes of Health, the National Science Foundation, Riverdale Country School, Character Lab, Facebook, and the John Templeton Foundation for their support. Of course, the views presented in this book are my own; they do not necessarily reflect the views of these organizations.

Finally, to Maya and Dani. The worst part about working on this book (by far) was knowing it took away from our time together. Thank you for your patience and love. I'm back!

Notes

Epigraphs

vii   **"The biggest challenge"**: Cathleen Falsani, "Transcript: Barack Obama and the God Factor Interview," *Sojourners*, March 27, 2012, sojo.net/articles/transcript-barack-obama-and-god-factor-interview.

vii   **"The voice in my head"**: Dan Harris, *10% Happier: How I Tamed the Voice in My Head, Reduced Stress Without Losing My Edge, and Found Self-Help That Actually Works—a True Story* (New York: It Books, 2014).

Introduction

xii   *CBS Evening News:* "Pain of Rejection: Real Pain for the Brain," CBS News, March 29, 2011, www.cbsnews.com/news/pain-of-rejection-real-pain-for-the-brain/. The segment can be viewed here: selfcontrol.psych.lsa.umich.edu/wp-content/uploads/2017/08/Why-does-a-broken-heart-physically-hurt.mp4.

xviii   **central evolutionary advances:** Janet Metcalfe and Hedy Kober, "Self-Reflective Consciousness and the Projectable Self," in *The Miss-*

ing Link in Cognition: Origins of Self-Reflective Consciousness, ed. H. S. Terrace and J. Metcalfe (Oxford: Oxford University Press, 2005), 57–83.

xviii **In recent years:** Each of the points referenced in this paragraph are fleshed out in the remaining chapters, with references provided when they are discussed. For a discussion of how chatter contributes to aging at the cellular level, see the "illnesses and infections" note in chapter 2.

xx *not* **living in the present:** Matthew A. Killingsworth and Daniel T. Gilbert, "A Wandering Mind Is an Unhappy Mind," *Science* 330 (2010): 932; Peter Felsman et al., "Being Present: Focusing on the Present Predicts Improvements in Life Satisfaction but Not Happiness," *Emotion* 17 (2007): 1047–1051; Michael J. Kane et al., "For Whom the Mind Wanders, and When, Varies Across Laboratory and Daily-Life Settings," *Psychological Science* 28 (2017): 1271–1289. As the Kane et al. article makes clear, mind wandering rates do, of course, vary across individuals. The numbers I report in the chapter refer to averages, like most of the other statistics I present in *Chatter.*

xx **"default state":** A paper published in 2001 triggered an explosion of research into the "default state," Marcus E. Raichle et al., "A Default Mode of Brain Function," *Proceedings of the National Academy of Sciences of the United States of America* 98 (2001): 676–682. Subsequent research linked default state activity to mind wandering: Malia F. Mason et al., "Wandering Minds: The Default Network and Stimulus-Independent Thought," *Science* 315 (2007): 393–395. Also see Kalina Christoff et al., "Experience Sampling During fMRI Reveals Default Network and Executive System Contributions to Mind Wandering," *Proceedings of the National Academy of Sciences of the United States of America* 106 (2009): 8719–8724.

xx **when we slip away:** As I explain in chapter 1, our default mode is not restricted to verbal reasoning. We can, for example, engage in visual-spatial reasoning when our mind wanders as well. Nonetheless, verbal reasoning constitutes a central component of mind-wandering. For example, in one of the first rigorous studies on this topic, Eric Klinger and W. Miles Cox concluded that "thought content is usually accompanied by some degree of interior monologue," which they defined as "I was talking to myself throughout the whole thought." They further noted that "interior monologues were at least as prevalent a feature of thought flow as visual imagery." Eric Klinger and W. Miles Cox, "Dimensions of Thought Flow in Everyday Life," *Imagination, Cognition, and Personality* 7 (1987): 105–128. Also see

Christopher L. Heavey and Russell T. Hurlburt, "The Phenomena of Inner Experience," *Consciousness and Cognition* 17 (2008): 798–810; and David Stawarczyk, Helena Cassol, and Arnaud D'Argembeau, "Phenomenology of Future-Oriented Mind-Wandering Episodes," *Frontiers in Psychology* 4 (2013): 1–12.

xx   **dawn of civilization:** Halvor Eifring, "Spontaneous Thought in Contemplative Traditions," in *The Oxford Handbook of Spontaneous Thought: Mind-Wandering, Creativity, and Dreaming,* ed. K. Christoff and K. C. R. Fox (New York: Oxford University Press, 2018), 529–538. Eifring conceptualizes spontaneous thought as a kind of mind-wandering, which, as noted above (see "when we slip away"), often involves interior monologue. More broadly, the idea that inner speech plays a prominent role in religion throughout history has been discussed by several scholars. Christopher C. H. Cook notes, for instance, "the attribution of voices to divine sources in contemporary religious experience is indisputable": Christopher C. H. Cook, *Hearing Voices, Demonic and Divine* (London: Routledge, 2019). For additional discussion, see Daniel B. Smith, *Muses, Madmen and Prophets: Hearing Voices and the Borders of Sanity* (New York: Penguin Books, 2007); T. M. Luhrmann, Howard Nusbaum, and Ronald Thisted, "The Absorption Hypothesis: Learning to Hear God in Evangelical Christianity," *American Anthropologist* 112 (2010): 66–78; Charles Fernyhough, *The Voices Within: The History and Science of How We Talk to Ourselves* (New York: Basic Books, 2016); and Douglas J. Davies, "Inner Speech and Religious Traditions," in *Theorizing Religion: Classical and Contemporary Debates,* ed. James A. Beckford and John Walliss (Aldershot, England: Ashgate Publishing, 2006), 211–223.

xxi   **one in ten people:** K. Maijer et al., "Auditory Hallucinations Across the Lifespan: A Systematic Review and Meta-Analysis," *Psychological Medicine* 48 (2018): 879–888.

xxi   **vocal impairments:** Ron Netsell and Klaas Bakker, "Fluent and Dysfluent Inner Speech of Persons Who Stutter: Self-Report," Missouri State University Unpublished Manuscript (2017). For discussion, see M. Perrone-Bertolotti et al., "What Is That Little Voice Inside My Head? Inner Speech Phenomenology, Its Role in Cognitive Performance, and Its Relation to Self-Monitoring," *Behavioural Brain Research* 261 (2014): 220–239, and Charles Fernyhough, *The Voices Within: The History and Science of How We Talk to Ourselves.* There is, however, evidence that people who stutter make errors during internal speech just as they do when they talk out loud when asked to perform tongue-twisters, "Investigating the Inner Speech of People

Who Stutter: Evidence for (and Against) the Covert Repair Hypothesis," *Journal of Communication Disorders* 44 (2011): 246–260.

xxi   **silently signing to themselves:** Deaf people who use sign language "talk to themselves" too, but the way their inner speech manifests shares both similarities and differences with hearing populations. Margaret Wilson and Karen Emmorey, "Working Memory for Sign Language: A Window into the Architecture of the Working Memory System," *Journal of Deaf Studies and Deaf Education* 2 (1997): 121–130; Perrone-Bertolotti et al., "What Is That Little Voice Inside My Head?"; and Helene Loevenbruck et al., "A Cognitive Neuroscience View of Inner Language: To Predict and to Hear, See, Feel," in *Inner Speech: New Voices,* ed. P. Langland-Hassan and Agustin Vicente (New York: Oxford University Press, 2019), 131–167. One brain imaging study found, for example, that the same regions of the left prefrontal cortex that supports inner speech in hearing populations becomes activated when profoundly deaf individuals were asked to silently complete a sentence (for example, "I am . . .") using inner signing. Philip K. McGuire et al., "Neural Correlates of Thinking in Sign Language," *NeuroReport* 8 (1997): 695–698. These findings are broadly consistent with research demonstrating an overlap between the brain systems that support spoken and signed language usage in hearing and deaf populations. To understand how signed and spoken language can share a common neural basis, it's useful to consider the fact that both types of languages are governed by *identical* sets of organizing principles (e.g., morphology, syntax, semantics, and phonology): Laura Ann Petitto et al., "Speech-Like Cerebral Activity in Profoundly Deaf People Processing Signed Languages: Implications for the Neural Basis of Human Language," *Proceedings of the National Academy of Sciences of the United States of America* 97 (2000): 13961–13966.

xxii   **four thousand words per minute:** Rodney J. Korba, "The Rate of Inner Speech," *Perceptual and Motor Skills* 71 (1990): 1043–1052, asked participants to record the "inner speech" they used to solve verbal word problems and then speak the solution out loud in fully predicated speech. Participants silently verbalized the solution approximately eleven times faster than they were able to express the solution in "expressive speech." As this study demonstrates, although we are capable of thinking to ourselves in full sentences, inner speech can also take a more condensed form that occurs much faster than how we talk out loud. For discussion, see Simon McCarthy Jones and Charles Fernyhough, "The Varieties of Inner Speech: Links Between

Quality of Inner Speech and Psychopathological Variables in a Sample of Young Adults," *Consciousness and Cognition* 20 (2011): 1586–1593.

xxii **State of the Union speeches:** I defined "contemporary American presidents' annual State of the Union speeches as referring to all presentations delivered from 2001 until the latest date that data was available in 2020. Gerhard Peters, "Length of State of the Union Address in Minutes (from 1966)," in The American Presidency Project, ed. John T. Woolley and Gerhard Peters (Santa Monica, CA: University of California, 1999–2020). Available from the World Wide Web: https://www.presidency.ucsb.edu/node/324136/.

xxii **sabotage us:** Psychologists have historically used different terms to refer to ostensibly similar chatter-related processes (for example, "rumination," "post-event processing," "habitual negative self-thinking," "chronic stress," and "worry"). Although in some cases subtle differences characterize these different forms of repetitive negative thinking (that is, rumination tends to be past focused, whereas worry is future oriented), scientists often talk about them as constituting a single construct of "perseverative cognition" or "negative repetitive thoughts." In this book, I use the term "chatter" to capture this concept. For discussion of these issues, see Jos F. Brosschot, William Gerin, and Julian F. Thayer, "The Perseverative Cognition Hypothesis: A Review of Worry, Prolonged Stress-Related Physiological Activation, and Health," *Journal of Psychosomatic Research* 60 (2006): 113–124; and Edward R. Watkins, "Constructive and Unconstructive Repetitive Thought," *Psychological Bulletin* 134 (2008): 163–206.

Chapter One: Why We Talk to Ourselves

3 **fourteen months:** For the date range of the project, see Irving's webpage at the University of Manchester: www.research.manchester.ac.uk/portal/en/researchers/andrew-irving(109e5208-716e-42e8-8d4f-578c9f556cd9)/projects.html?period=finished.

4 **a hundred New Yorkers:** "Interview: Dr. Andrew Irving & 'New York Stories,'" June 10, 2013, Wenner-Gren Foundation, blog .wennergren.org/2013/06/interview-dr-andrew-irving-new-york-stories/; and Andrew Irving, *The Art of Life and Death: Radical Aesthetics and Ethnographic Practice* (New York: Hau Books, 2017).

4   **earlier fieldwork in Africa:** For a discussion of Irving's fieldwork in Africa see Andrew Irving, "Strange Distance: Towards an Anthropology of Interior Dialogue," *Medical Anthropology Quarterly* 25 (2011): 22–44; and Sydney Brownstone, "For 'New York Stories,' Anthropologist Tracked 100 New Yorkers' Inner Monologues Across the City," *Village Voice,* May 1, 2013.

7   **avid time traveler:** Thomas Suddendorf and Michael C. Corballis, "The Evolution of Foresight: What Is Mental Time Travel, and Is It Unique to Humans?," *Behavioral and Brain Sciences* 30 (2007): 299–351.

8   **often dealt with negative "content":** Irving noted that although there was variability in what participants thought about, he was struck by how many people thought about negative topics such as economic instability and terrorism. Brownstone, "For 'New York Stories,' Anthropologist Tracked 100 New Yorkers' Inner Monologues Across the City."

8   **nature of the default state:** Eric Klinger, Ernst H. W. Koster, and Igor Marchetti, "Spontaneous Thought and Goal Pursuit: From Functions Such as Planning to Dysfunctions Such as Rumination," in Christoff and Fox, *Oxford Handbook of Spontaneous Thought,* 215–232; Arnaud D'Argembeau, "Mind-Wandering and Self-Referential Thought," in ibid., 181–192; and A. Morin, B. Uttl, and B. Hamper, "Self-Reported Frequency, Content, and Functions of Inner Speech," *Procedia: Social and Behavioral Journal* 30 (2011): 1714–1718.

9   **nonverbal forms:** See "when we slip away" note from the introduction.

9   **neural reuse:** Michael L. Anderson, "Neural Reuse: A Fundamental Principle of the Brain," *Behavioral and Brain Sciences* 33 (2010): 245–313.

11   **phonological loop:** Alan Baddeley, "Working Memory," *Science* 255 (1992): 556–559. Also see Alan Baddeley and Vivien Lewis, "Inner Active Processes in Reading: The Inner Voice, the Inner Ear, and the Inner Eye," in *Interactive Processes in Reading,* ed. A. M. Lesgold and C. A. Perfetti (Hillsdale, NJ: Lawrence Erlbaum, 1981), 107–129; Alan D. Baddeley and Graham J. Hitch, "The Phonological Loop as a Buffer Store: An Update," *Cortex* 112 (2019): 91–106; and Antonio Chella and Arianna Pipitone, "A Cognitive Architecture for Inner Speech," *Cognitive Systems Research* 59 (2020): 287–292.

11   **in infancy:** Nivedita Mani and Kim Plunkett, "In the Infant's Mind's Ear: Evidence for Implicit Naming in 18-Month-Olds," *Psychological*

Science 21 (2010): 908–913. For discussion, see Ben Alderson-Day and Charles Fernyhough, "Inner Speech: Development, Cognitive Functions, Phenomenology, and Neurobiology," *Psychological Bulletin* 141 (2015); and Perrone-Bertolotti et al., "What Is That Little Voice Inside My Head?"

11   **language development and self-control:** Lev Vygotsky, *Thinking and Speech: The Collected Works of Lev Vygotsky,* vol. 1 (1934; New York: Plenum Press, 1987). Also see Alderson-Day and Fernyhough, "Inner Speech"; and Perrone-Bertolotti et al., "What Is That Little Voice Inside My Head?"

12   **research on socialization:** For research highlighting the complexity of the role that parents play in socialization, see W. Andrew Collins et al., "Contemporary Research on Parenting: The Case for Nature and Nurture," *American Psychologist* 55 (2000): 218–232. A more recent illustration of the role that parents play in children's emotional lives comes from a large meta-analysis, which revealed statistically significant positive links between parental behavior and several emotional adjustment outcomes. See Michael M. Barger et al., "The Relation Between Parents' Involvement in Children's Schooling and Children's Adjustment: A Meta-analysis," *Psychological Bulletin* 145 (2019): 855–890.

12   **shape our own verbal streams:** For broader discussions of the role that language plays in the transmission of cultural ideas, see Susan A. Gelman and Steven O. Roberts, "How Language Shapes the Cultural Inheritance of Categories," *Proceedings of the National Academy of Sciences of the United States of America* 114 (2017): 7900–7907; and Roy Baumeister and E. J. C. Masicampo, "Conscious Thought Is for Facilitating Social and Cultural Interactions," *Psychological Review* 117 (2010): 945–971.

12   **broader cultural factors:** Hazel R. Markus and Shinobu Kitayama, "Culture and the Self: Implications for Cognition, Emotion, and Motivation," *Psychological Review* 98 (1991): 224–253.

12   **Religions and the values they teach:** Adam B. Cohen, "Many Forms of Culture," *American Psychologist* 64 (2009): 194–204.

13   **inner speech earlier:** Laura E. Berk and Ruth A. Garvin, "Development of Private Speech Among Low-Income Appalachian Children," *Developmental Psychology* 20 (1984): 271–286; Laura E. Berk, "Children's Private Speech: An Overview of Theory and the Status of Research," in *Private Speech: From Social Interaction to Self-Regulation,* eds.

Rafael M. Diaz and Laura E. Berk (New York: Psychology Press, 1992), 17–54.

13   **imaginary friends may spur internal speech:** Paige E. Davis, Elizabeth Meins, and Charles Fernyhough, "Individual Differences in Children's Private Speech: The Role of Imaginary Companions," *Journal of Experimental Child Psychology* 116 (2013): 561–571.

13   **among many other desirable qualities:** Amanda Grenell and Stephanie M. Carlson, "Pretense," in *The Sage Encyclopedia of Contemporary Early Childhood Education,* ed. D. Couchenour and J. K. Chrisman (New York: Sage, 2016), 1075–1077.

13   **spontaneous thoughts related to goals:** For illustrative studies, see Arnaud D'Argembeau, Olivier Renaud, and Martial Van der Linden, "Frequency, Characteristics, and Functions of Future-Oriented Thoughts in Daily Life," *Applied Cognitive Psychology* 25 (2011): 96–103; Alain Morin, Christina Duhnych, and Famira Racy, "Self-Reported Inner Speech Use in University Students," *Applied Cognitive Psychology* 32 (2018): 376–382; and Akira Miyake et al., "Inner Speech as a Retrieval Aid for Task Goals: The Effects of Cue Type in the Random Task Cuing Paradigm," *Acta Psychologica* 115 (2004): 123–142. Also see Adam Winsler, "Still Talking to Ourselves After All These Years: A Review of Current Research on Private Speech," in *Private Speech, Executive Functioning, and the Development of Verbal Self-Regulation,* ed. A. Winsler, C. Fernyhough, and I. Montero (New York: Cambridge University Press, 2009), 3–41.

14   **run mental simulations:** D'Argembeau, Renaud, and Van der Linden, "Frequency, Characteristics, and Functions of Future-Oriented Thoughts in Daily Life"; D'Argembeau, "Mind-Wandering and Self-Referential Thought"; and Morin, Duhnych, and Racy, "Self-Reported Inner Speech Use in University Students."

14   **Historically, psychologists thought of dreams:** Erin J. Wamsley, "Dreaming and Waking Thought as a Reflection of Memory Consolidation," in Christoff and Fox, *Oxford Handbook of Spontaneous Thought,* 457–468, presents a cogent review of dream research.

15   **share many similarities:** Kieran C. R. Fox et al., "Dreaming as Mind Wandering: Evidence from Functional Neuroimaging and First-Person Content Reports," *Frontiers in Human Neuroscience* 7 (2013): 1–18; Tracey L. Kahan and Stephen P. LaBerge, "Dreaming and Waking: Similarities and Differences Revisited," *Consciousness and Cognition* 20 (2011): 494–514; Lampros Perogamvros et al., "The Phenomenal Contents and Neural Correlates of Spontaneous

Thoughts Across Wakefulness, NREM Sleep, and REM Sleep," *Journal of Cognitive Neuroscience* 29 (2017): 1766–1777; and Erin J. Wamsley, "Dreaming and Waking Thought as a Reflection of Memory Consolidation."

15 **dreams are often functional:** For a discussion of the role that dreams play in simulating threats, see Katja Valli and Antti Revonsuo, "The Threat Simulation Theory in Light of Recent Empirical Evidence: A Review," *American Journal of Psychology* 122 (2009): 17–38; and Antti Revonsuo, "The Reinterpretation of Dreams: An Evolutionary Hypothesis of the Function of Dreaming," *Behavioral and Brain Sciences* 23 (2001): 877–901. Also see J. Allan Hobson, "REM Sleep and Dreaming: Towards a Theory of Protoconsciousness," *Nature Reviews Neuroscience* 10 (2009): 803–813.

15 **creation of our selves:** Arnaud D'Argembeau et al., "Brains Creating Stories of Selves: The Neural Basis of Autobiographical Reasoning," *Social Cognitive Affective Neuroscience* 9 (2014): 646–652; Raymond A. Mar, "The Neuropsychology of Narrative: Story Comprehension, Story Production, and Their Interrelation," *Neuropsychologia* 42 (2004): 1414–1434; and Baumeister and Masicampo, "Conscious Thought Is for Facilitating Social and Cultural Interactions"; Kate C. McLean et al., "Selves Creating Stories Creating Selves: A Process Model of Self-Development," *Personality and Social Psychology Review* 11 (2007): 262–278. For a broader discussion of the role that language plays in autobiographical reasoning, see Robyn Fivus, "The Stories We Tell: How Language Shapes Autobiography," *Applied Cognitive Psychology* 12 (1998): 483–487.

16 **stopped functioning well:** To tell Jill Bolte Taylor's story, I drew on her book, *My Stroke of Insight: A Brain Scientist's Personal Journey* (New York: Penguin Books, 2008), and her TED Talk, "My Stroke of Insight," www.ted.com/talks/jill_bolte_taylor_s_powerful_stroke_of_insight?language=en, both of which I quote from. I am grateful to an article by Alain Morin that analyzed Jill Bolte Taylor's case in the context of private speech for pointing me to this example: Alain Morin, "Self-Awareness Deficits Following Loss of Inner Speech: Dr. Jill Bolte Taylor's Case Study," *Consciousness and Cognition* 18 (2009): 524–529.

19 **inner experiences consistently dwarf outer ones:** Killingsworth and Gilbert, "Wandering Mind Is an Unhappy Mind."

Chapter Two: When Talking to Ourselves Backfires

21   **first wild pitch:** To tell Rick Ankiel's story, I drew on Rick Ankiel, *The Phenomenon: Pressure, the Yips, and the Pitch That Changed My Life* (New York: PublicAffairs, 2017), which I quote from, as well as this article: Gary Waleik, "Former MLB Hurler Remembers 5 Pitches That Derailed His Career," *Only a Game*, WBUR, May 19, 2017, www.wbur.org/onlyagame/2017/05/19/rick-ankiel-baseball; and Rick Ankiel, "Letter to My Younger Self," *The Players' Tribune*, Sept. 18, 2017, https://www.theplayerstribune.com/en-us/articles/rick-ankiel -letter-to-my-younger-self-cardinals.

22   *national TV*: Waleik, "Former MLB Hurler Remembers 5 Pitches That Derailed His Career."

22   **crowd oohed a bit louder:** MLB.com. YouTube: https://www.you tube.com/watch?time_continue=5&v=KDZX525CSvw&feature= emb_title.

24   **never pitch professionally again:** Baseball-reference.com: https:// www.baseball-reference.com/players/a/ankieri01.shtml.

24   **influence our** *attention:* Sian Beilock is one of the world's foremost experts on choking under pressure. I drew on the work she describes in Sian L. Beilock and Rob Gray, "Why Do Athletes Choke Under Pressure?," in *Handbook of Sport Psychology*, 3rd ed., ed. G. Tenenbaum and R. C. Eklund (Hoboken, NJ: John Wiley and Sons, 2007), 425– 444.

25   **Attention is what allows us:** Michael I. Posner and Mary K. Roth- bart, "Research on Attention Networks as a Model for the Integra- tion of Psychological Science," *Annual Review of Psychology* 58 (2007): 1–23.

25   **"It's a move that requires":** Amanda Prahl, "Simone Biles Made History with Her Triple Double—Here's What That Term Actually Means," *PopSugar*, Aug. 15, 2019, www.popsugar.com/fitness/What -Is-Triple-Double-in-Gymnastics-46501483. Also see Charlotte Caroll, "Simone Biles Is First-Ever Woman to Land Triple Double in Competition on Floor," *Sports Illustrated*, Aug. 11, 2019, https:// www.si.com/olympics/2019/08/12/simone-biles-first-ever-woman -land-triple-double-competition-video.

26   **He** *unlinked:* Beilock and Gray, "Why Do Athletes Choke Under Pressure?" Note that this work typically uses the word "dechunked" to describe the process that I refer to as "unlinked."

27   **paralysis by analysis:** Sian Beilock, *Choke* (New York: Little, Brown, 2011).

27   **steer our thoughts and behavior:** Adele Diamond, "Executive Functions," *Annual Review of Psychology* 64 (2013): 135–168.

28   **limited capacity:** Amitai Shenhav et al., "Toward a Rational and Mechanistic Account of Mental Effort," *Annual Review of Neuroscience* 40 (2017): 99–124.

28   **illustration of this limited capacity:** Nelson Cowan, "The Magical Mystery Four: How Is Working Memory Capacity Limited, and Why?," *Current Directions in Psychological Science* 19 (2010): 51–57.

28   **hogs our neural capacity:** The idea that perseverative cognition compromises executive functions has been studied from several perspectives. See Michael W. Eysenck et al., "Anxiety and Cognitive Performance: Attentional Control Theory," *Emotion* 7 (2007): 336–353; Hannah R. Snyder, "Major Depressive Disorder Is Associated with Broad Impairments on Neuropsychological Measures of Executive Function: A Meta-analysis and Review," *Psychological Bulletin* 139 (2013): 81–132; and Tim P. Moran, "Anxiety and Working Memory Capacity: A Meta-analysis and Narrative Review," *Psychological Bulletin* 142 (2016): 831–864.

29   **perform worse on tests:** Nathaniel von der Embse et al., "Test Anxiety Effects, Predictors, and Correlates: A 30-Year Meta-analytic Review," *Journal of Affective Disorders* 227 (2018): 483–493.

29   **artistic performers:** Dianna T. Kenny, "A Systematic Review of Treatments for Music Performance Anxiety," *Anxiety, Stress, and Coping* 18 (2005): 183–208.

29   **make low initial offers:** Alison Wood Brooks and Maurice E. Schweitzer, "Can Nervous Nelly Negotiate? How Anxiety Causes Negotiators to Make Low First Offers, Exit Early, and Earn Less Profit," *Organizational Behavior and Human Decision Processes* 115 (2011): 43–54.

30   **Bernard Rimé:** Bernard Rimé, "Emotion Elicits the Social Sharing of Emotion: Theory and Empirical Review," *Emotion Review* 1 (2009): 60–85. I also drew on the following lecture: Bernard Rimé, "The Social Sharing of Emotion" (lecture delivered at Collective Emotions in Cyberspace Consortium), YouTube, published May 20, 2013, www.youtube.com/watch?v=JdCksLisfUQ.

30   **From Asia to the Americas:** Although Rimé's research suggests that the motivation to talk about one's emotions is a cross-cultural phe-

nomenon, cultures nonetheless vary in the rate at which they share their emotions. See Archana Singh-Manoux and Catrin Finkenauer, "Cultural Variations in Social Sharing of Emotions: An Intercultural Perspective on a Universal Phenomenon," *Journal of Cross-Cultural Psychology* 32 (2001): 647–661. Also see Heejung S. Kim, "Social Sharing of Emotion in Words and Otherwise," *Emotion Review* 1 (2009): 92–93.

31   **pushing away:** For review, see Susan Nolen-Hoeksema, Blair E. Wisco, and Sonja Lyubomirsky, "Rethinking Rumination," *Perspectives on Psychological Science* 3 (2008): 400–424; also see Thomas E. Joiner et al., "Depression and Excessive Reassurance-Seeking," *Psychological Inquiry* 10 (1999): 269–278; Michael B. Gurtman, "Depressive Affect and Disclosures as Factors in Interpersonal Rejection," *Cognitive Therapy Research* 11 (1987): 87–99; and Jennifer L. Schwartz and Amanda McCombs Thomas, "Perceptions of Coping Responses Exhibited in Depressed Males and Females," *Journal of Social Behavior and Personality* 10 (1995): 849–860.

31   **less capable of solving problems:** For reviews, see Nolen-Hoeksema, Wisco, and Lyubomirsky, "Rethinking Rumination"; and Lyubomirsky et al., "Thinking About Rumination," *Annual Review of Clinical Psychology* 11 (2015): 1–22.

31   **toxic outcome:** For a discussion of how frayed social relationships contribute to feelings of social isolation and loneliness, see Julianne Holt-Lunstad, "Why Social Relationships Are Important for Physical Health: A Systems Approach to Understanding and Modifying Risk and Perception," *Annual Review of Psychology* 69 (2018): 437–458; and Julianne Holt-Lunstad, Timothy B. Smith, Mark Baker, Tyler Harris, and David Stephenson, "Loneliness and Social Isolation as Risk Factors for Mortality: A Meta-analytic Review," *Perspectives on Psychological Science* 10 (2015): 227–237.

For work documenting the toxic effects of loneliness and social isolation, see John T. Cacioppo and Stephanie Cacioppo, "The Growing Problem of Loneliness," *The Lancet* 391 (2018): 426; Greg Miller, "Why Loneliness Is Hazardous to Your Health," *Science* 14 (2011): 138–140; and Aparna Shankar, Anne McMunn, James Banks, and Andrew Steptoe, "Loneliness, Social Isolation, and Behavioral and Biological Health Indicators in Older Adults," *Health Psychology* 30 (2011): 377–385.

32   **kids who were prone to rumination:** Katie A. McLaughlin and Susan Nolen-Hoeksema, "Interpersonal Stress Generation as a Mech-

anism Linking Rumination to Internalizing Symptoms in Early Adolescents," *Journal of Clinical Child and Adolescent Psychology* 41 (2012): 584–597.

A study by John Cacioppo and colleagues further underscores the reciprocal link between loneliness and self-focused attention: John T. Cacioppo, Hsi Yuan Chen, and Stephanie Cacioppo, "Reciprocal Influences Between Loneliness and Self-Centeredness: A Cross-Lagged Panel Analysis in a Population-Based Sample of African American, Hispanic, and Caucasian Adults," *Personality and Social Psychology Bulletin* 43 (2017): 1125–1135.

32 **grieving adults:** Susan Nolen-Hoeksema and Christopher G. Davis, " 'Thanks for Sharing That': Ruminators and Their Social Support Networks," *Journal of Personality and Social Psychology* 77 (1999): 801–814.

32 **behave aggressively:** Thomas F. Denson et al., "Understanding Impulsive Aggression: Angry Rumination and Reduced Self-Control Capacity Are Mechanisms Underlying the Provocation-Aggression Relationships," *Personality and Social Psychology Bulletin* 37 (2011): 850–862; and Brad J. Bushman, "Does Venting Anger Feed or Extinguish the Flame? Catharsis, Rumination, Distraction, Anger, and Aggressive Responding," *Personality and Social Psychology Bulletin* 28 (2002): 724–731.

32 **displace our aggression:** Brad J. Bushman et al., "Chewing on It Can Chew You Up: Effects of Rumination on Triggered Displaced Aggression," *Journal of Personality and Social Psychology* 88 (2005): 969–983.

33 **two and a half billion people:** Facebook Newsroom, Facebook, newsroom.fb.com/company-info/; and J. Clement, "Number of Monthly Active Twitter Users Worldwide from 1st Quarter 2010 to 1st Quarter 2019 (in Millions)," Statista, www.statista.com/statistics/282087/number-of-monthly-active-twitter-users/.

33 **share their private ruminations:** Mina Choi and Catalina L. Toma, "Social Sharing Through Interpersonal Media: Patterns and Effects on Emotional Well-Being," *Computers in Human Behavior* 36 (2014): 530–541; and Adriana M. Manago, Tamara Taylor, and Patricia M. Greenfield, "Me and My 400 Friends: The Anatomy of College Students' Facebook Networks, Their Communication Patterns, and Well-Being," *Developmental Psychology* 48 (2012): 369–380.

33 **how we interact with them:** As one example of this principle, consider research my colleagues and I performed demonstrating that pas-

sively using Facebook (that is, browsing the site to consume information about others) leads to emotional well-being declines, whereas actively using Facebook (that is, producing information on the site) does not. See Philippe Verduyn et al., "Passive Facebook Usage Undermines Affective Well-Being: Experimental and Longitudinal Evidence," *Journal of Experimental Psychology: General* 144 (2015): 480–488. For review, see Philippe Verduyn et al., "Do Social Network Sites Enhance or Undermine Subjective Well-Being? A Critical Review," *Social Issues and Policy Review* 11 (2017): 274–302.

33   **importance of empathy:** Jamil Zaki, *The War for Kindness: Building Empathy in a Fractured World* (New York: Crown, 2019); and Frans B. M. de Waal and Stephanie Preston, "Mammalian Empathy: Behavioural Manifestations and Neural Basis," *Nature Reviews Neuroscience* 18 (2017): 498–509.

33   **find ourselves venting:** Rimé, "Emotion Elicits the Social Sharing of Emotion."

34   **subtle physical gestures:** John Suler, "The Online Disinhibition Effect," *Cyberpsychology and Behavior* 3 (2004): 321–326; Noam Lapidot-Lefler and Azy Barak, "Effects of Anonymity, Invisibility, and Lack of Eye-Contact on Toxic Online Disinhibition," *Computers in Human Behavior* 28 (2012): 434–443; and Christopher Terry and Jeff Cain, "The Emerging Issue of Digital Empathy," *American Journal of Pharmaceutical Education* 80 (2016): 58.

34   **Cyberbullying:** Committee on the Biological and Psychosocial Effects of Peer Victimization: Lessons for Bullying Prevention, National Academy of Sciences Report; Michele P. Hamm et al., "Prevalence and Effect of Cyberbullying on Children and Young People," *JAMA Pediatrics,* Aug. 2015; Robin M. Kowalski et al., "Bullying in the Digital Age: A Critical Review and Meta-analysis of Cyberbullying Research Among Youth," *Psychological Bulletin* 140 (2014): 1073–1137; and Robert Tokunaga, "Following You Home from School: A Critical Review and Synthesis of Research on Cyberbullying Victimization," *Computers in Human Behavior* 26 (2010): 277–287.

34   **passage of time:** Emotions typically wane once they reach their maximum level of intensity: Philippe Verduyn, Iven Van Mechelen, and Francis Tuerlinckx, "The Relation Between Event Processing and the Duration of Emotional Experience," *Emotion* 11 (2011): 20–28; and Philippe Verduyn et al., "Predicting the Duration of Emo-

tional Experience: Two Experience Sampling Studies," *Emotion* 9 (2009): 83–91.

35 **irritate and alienate others:** Caitlin McLaughlin and Jessica Vitak, "Norm Evolution and Violation on Facebook," *New Media and Society* 14 (2012): 299–315; and Emily M. Buehler, "'You Shouldn't Use Facebook for That': Navigating Norm Violations While Seeking Emotional Support on Facebook," *Social Media and Society* 3 (2017): 1–11.

35 **share more negative personal content:** Jiyoung Park et al., "When Perceptions Defy Reality: The Relationships Between Depression and Actual and Perceived Facebook Social Support," *Journal of Affective Disorders* 200 (2016): 37–44.

35 **need to self-present:** For two classic accounts of the role that self-presentation plays in daily life, see Erving Goffman, *The Presentation of Self in Everyday Life* (Garden City, NY: Doubleday, 1959); and Mark R. Leary and Robin M. Kowalski, "Impression Management: A Literature Review and Two-Component Model," *Psychological Bulletin* 107 (1990): 34–47.

35 **skillfully curate:** Randi Zuckerberg captured this facet of Facebook well in an interview she did with *The New York Times*. "What are you most guilty of on Facebook?" the reporter asked her. "I'm a marketer," she responded, "and sometimes I almost can't take it out of my personal life. I've had friends call me and say, 'Your life looks so amazing.' And I tell them: 'I'm a marketer; I'm only posting the moments that are amazing.'" Susan Dominus, "Randi Zuckerberg: 'I Really Put Myself Out There,'" *New York Times,* Nov. 1, 2013, www.nytimes.com/2013/11/03/magazine/randi-zuckerberg-i-really -put-myself-out-there.html.

35 **feel better:** Amy L. Gonzales and Jeffrey T. Hancock, "Mirror, Mirror on My Facebook Wall: Effects of Exposure to Facebook on Self-Esteem," *Cyberpsychology, Behavior, and Social Networking* 14 (2011): 79–83.

36 **driven to compare ourselves with others:** Leon Festinger, "A Theory of Social Comparison Processes," *Human Relations* 7 (1954): 117–140; and Katja Corcoran, Jan Crusius, and Thomas Mussweiler, "Social Comparison: Motives, Standards, and Mechanisms," in *Theories in Social Psychology,* ed. D. Chadee (Oxford: Wiley-Blackwell, 2011), 119–139. Sometimes we compare ourselves with others to see how we're stacking up in a particular domain. Other times it's to

make ourselves feel better (by comparing ourselves with someone ostensibly "beneath" us) or to identify how we might improve some facet of our lives that we care about (by comparing ourselves with someone ostensibly "above" us). There is also evidence that comparing ourselves with others is an efficient way of measuring and obtaining information about ourselves.

36   **A study my colleagues and I published:** Verduyn et al., "Passive Facebook Usage Undermines Affective Well-Being."

And the more we stew over how badly our lives stack up against others, the worse the consequences. Case in point: A longitudinal study performed with 268 young adults found that the more people compared themselves negatively to others on Facebook, the more they ruminated and the more depressed they felt: Feinstein et al., "Negative Social Comparison on Facebook and Depressive Symptoms," *Psychology of Popular Media Culture* 2 (2013): 161–170.

Also see Melissa G. Hunt et al., "No More FOMO: Limiting Social Media Decreases Loneliness and Depression," *Journal of Social and Clinical Psychology* 37 (2018): 751–768; Morten Tromholt, "The Facebook Experiment: Quitting Facebook Leads to Higher Levels of Well-Being," *Cyberpsychology, Behavior, and Social Networking* 19 (2016): 661–666; R. Mosquera et al., "The Economic Effects of Facebook," *Experimental Economics* (2019); Holly B. Shakya and Nicholas A. Christakis, "Association of Facebook Use with Compromised Well-Being: A Longitudinal Study," *American Journal of Epidemiology* 185 (2017): 203–211; and Cesar G. Escobar-Viera et al., "Passive and Active Social Media Use and Depressive Symptoms Among United States Adults," *Cyberpsychology, Behavior, and Social Networking* 21 (2018): 437–443.

Research has also begun to demonstrate how these findings generalize to other social media platforms like Instagram. Eline Frison and Steven Eggermont, "Browsing, Posting, and Liking on Instagram: The Reciprocal Relationships Between Different Types of Instagram Use and Adolescents' Depressed Mood," *Cyberpsychology, Behavior, and Social Networking* 20 (2017): 603–609.

36   **the more envy they experienced:** The negative consequences of envy are well-established. However, envy isn't all bad. It can be functional in small doses, motivating us to improve ourselves: Jens Lange, Aaron Weidman, and Jan Crusius, "The Painful Duality of Envy: Evidence for an Integrative Theory and a Meta-analysis on the Relation of Envy and Schadenfreude," *Journal of Personality and Social Psychology* 114 (2018): 572–598.

36   **One answer to that question:** Additional explanations for why we continue to use social media in spite of its negative consequences include: (a) our desire to stay abreast of what is happening in our community, which might trump our desire to feel better about ourselves at any given moment in time, (b) the desire to obtain feedback from others, and (c) people often misjudge how using Facebook will make them feel (i.e., we focus on the potential positives that social media will bring us, losing sight [or perhaps even being unaware in the first place] of its potential to do harm as well). For discussion, see Ethan Kross and Susannah Cazaubon, "How Does Social Media Influence People's Emotional Lives?," in *Applications of Social Psychology: How Social Psychology Can Contribute to the Solution of Real-World Problems,* eds. J. Forgas, William D. Crano, and Klaus Fiedler (New York: Routledge-Psychology Press, 2020), 250–264.

36   **Harvard neuroscientists:** Diana I. Tamir and Jason P. Mitchell, "Disclosing Information About the Self Is Intrinsically Rewarding," *Proceedings of the National Academy of Sciences of the United States of America* 109 (2012): 8038–8043.

37   **languages across the globe:** Geoff MacDonald and Mark R. Leary, "Why Does Social Exclusion Hurt? The Relationship Between Social and Physical Pain," *Psychological Bulletin* 131 (2005): 202–223; Naomi I. Eisenberger, Matthew D. Lieberman, and Kipling D. Williams, "Does Rejection Hurt? An fMRI Study of Social Exclusion," *Science* 302 (2003): 290–292.

37   **heartbroken of New York City:** Ethan Kross et al., "Social Rejection Shares Somatosensory Representations with Physical Pain," *Proceedings of the National Academy of Sciences of the United States of America* 108 (2011): 6270–6275.

38   **city of eight million:** https://www.health.ny.gov/statistics/vital_statistics/2007/table02.htm.

38   **influence what happens in our bodies:** Naomi I. Eisenberger and Steve W. Cole, "Social Neuroscience and Health: Neurophysiological Mechanisms Linking Social Ties with Physical Health," *Nature Neuroscience* 15 (2012): 669–674; and Gregory Miller, Edith Chen, and Steve W. Cole, "Health Psychology: Developing Biologically Plausible Models Linking the Social World and Physical Health," *Annual Review of Psychology* 60 (2009): 501–524.

38   **$500 billion annually:** Michele Hellebuyck et al., "Workplace Health Survey," Mental Health America, www.mhanational.org

/sites/default/files/Mind%20the%20Workplace%20-%20MHA%20
Workplace%20Health%20Survey%202017%20FINAL.pdf.

39   **negative verbal stream:** For an account of how perseverative cognition, which often takes the form of verbal rumination and worry (see
the introduction), prolongs the stress response, see Brosschot, Gerin,
and Thayer, "Perseverative Cognition Hypothesis"; Jos F. Brosschot,
"Markers of Chronic Stress: Prolonged Physiological Activation and
(Un)conscious Perseverative Cognition," *Neuroscience and Biobehavioral
Reviews* 35 (2010): 46–50; and Cristina Ottaviani et al., "Physiological Concomitants of Perseverative Cognition: A Systematic Review
and Meta-analysis," *Psychological Bulletin* 142 (2016): 231–259.

40   **illnesses that span the gamut:** Andrew Steptoe and Mika Kivimaki,
"Stress and Cardiovascular Disease," *Nature Reviews Cardiology* 9
(2012): 360–370; Suzanne C. Segerstrom and Gregory E. Miller,
"Psychological Stress and the Human Immune System: A Meta-
analytic Study of 30 Years of Inquiry," *Psychological Bulletin* 130
(2004): 601–630; Bruce S. McEwen, "Brain on Stress: How the Social Environment Gets Under the Skin," *Proceedings of the National
Academy of Sciences of the United States of America* 109 (2012): 17180–
17185; Ronald Glaser and Janice Kiecolt-Glaser, "Stress-Induced Im-
mune Dysfunction: Implications for Health," *Nature Reviews
Immunology* 5 (2005): 243–251; Edna Maria Vissoci Reiche, Sandra
Odebrecht Vargas Nunes, and Helena Kaminami Morimoto, "Stress,
Depression, the Immune System, and Cancer," *Lancet Oncology* 5
(2004): 617–625; A. Janet Tomiyama, "Stress and Obesity," *Annual
Review of Psychology* 70 (2019): 703–718; and Gregory E. Miller et al.,
"A Functional Genomic Fingerprint of Chronic Stress in Humans:
Blunted Glucocorticoid and Increased NF-κB Signaling," *Biological
Psychiatry* 15 (2008): 266–272.

40   **not having a strong social-support network:** Julianne Holt-
Lunstad, Timothy B. Smith, and J. Bradley Layton, "Social Relationships and Mortality Risk: A Meta-analytic Review," *PLOS Medicine*
7 (2010): e1000316.

40   ***transdiagnostic risk factor:*** Susan Nolen-Hoeksema and Edward R.
Watkins, "A Heuristic for Developing Transdiagnostic Models of
Psychopathology: Explaining Multifinality and Divergent Trajecto-
ries," *Perspectives on Psychological Science* 6 (2011): 589–609; Katie A.
McLaughlin et al., "Rumination as a Transdiagnostic Factor Under-
lying Transitions Between Internalizing Symptoms and Aggressive
Behavior in Early Adolescents," *Journal of Abnormal Psychology* 123

(2014): 13–23; Edward R. Watkins, "Depressive Rumination and Co-morbidity: Evidence for Brooding as a Transdiagnostic Process," *Journal of Rational-Emotive and Cognitive-Behavior Therapy* 27 (2009): 160–75; Douglas S. Mennin and David M. Fresco, "What, Me Worry and Ruminate About DSM-5 and RDoC? The Importance of Targeting Negative Self-Referential Processing," *Clinical Psychology: Science and Practice* 20 (2013): 258–267; and Brosschot, "Markers of Chronic Stress."

41   **DNA is like a piano:** I drew on the following sources to make the connection between gene expression and playing a musical instrument: Jane Qiu, "Unfinished Symphony," *Nature* 441 (2006): 143–145; and University of Texas Health Science Center at San Antonio, "Study Gives Clue as to How Notes Are Played on the Genetic Piano," *EurekAlert!*, May 12, 2011, www.eurekalert.org/pub _releases/2011-05/uoth-sgc051011.php.

41   **Steve Cole:** Steven W. Cole, "Social Regulation of Human Gene Expression," *American Journal of Public Health* 103 (2013): S84–S92. I also drew on the following talk that Steve delivered at Stanford: "Meng-Wu Lecture" (lecture delivered at the Center for Compassion and Altruism Research and Education, Nov. 12, 2013), ccare.stanford .edu/videos/meng-wu-lecture-steve-cole-ph-d/.

41   **inflammation genes:** George M. Slavich and Michael R. Irwin, "From Stress to Inflammation and Major Depressive Disorder: A Social Signal Transduction Theory of Depression," *Psychological Bulletin* 140 (2014): 774–815; Steve W. Cole et al., "Social Regulation of Gene Expression in Human Leukocytes," *Genome Biology* 8 (2007): R189; and Gregory E. Miller, Edith Chen, and Karen J. Parker, "Psychological Stress in Childhood and Susceptibility to the Chronic Diseases of Aging: Moving Towards a Model of Behavioral and Biological Mechanisms," *Psychological Bulletin* 137 (2011): 959–997.

42   **illnesses and infections:** Chatter also extends its tentacles around our DNA in another fashion—through our *telomeres*. Telomeres are little caps at the end of our chromosomes that protect our DNA from unraveling in ways that affect our health and longevity. Short telomeres contribute to a host of age-related diseases. Fortunately, we all have a chemical in our body called *telomerase* that is capable of preserving the length of our telomeres. The problem is, stress hormones like cortisol deplete our body of this chemical, speeding up the rate at which our telomeres shorten.

In 2004, Elissa Epel, Nobel laureate Elizabeth Blackburn, and their colleagues published a landmark study examining the relationship between how stressed women felt over a ten-month period and their telomere length. As expected, they found that the more stressed the women felt—stress, of course, being a trigger for chatter, and chatter a driver of *chronic* stress—the shorter their telomeres. Even more dramatic, the most stressed women had telomeres that were equivalent to *over a decade shorter* than the least stressed women. Elissa S. Epel et al., "Accelerated Telomere Shortening in Response to Life Stress," *Proceedings of the National Academy of Sciences* 101 (2004): 17312–17315.

For a detailed review, see Elizabeth H. Blackburn and Elissa S. Epel, *The Telomere Effect* (New York: Grand Central Publishing, 2017). Also see Elizabeth Blackburn, Elissa S. Epel, and Jue Lin, "Human Telomere Biology: A Contributory and Interactive Factor in Aging, Disease Risks, and Protection," *Science* 350 (2015): 1193–1198; and Kelly E. Rentscher et al., "Psychosocial Stressors and Telomere Length: A Current Review of the Science," *Annual Review of Public Health* 41 (2020): 223–245.

43   **nearly twenty years:** Matt Kelly, "This Thirty-Nine-Year-Old Is Attempting a Comeback," MLB.com, August 2, 2018, https://www.mlb.com/news/rick-ankiel-to-attempt-comeback-c288544452 (retrieved February 9, 2020).

Chapter Three: Zooming Out

44   **"Have you ever killed someone?":** I changed the name and several other details in this story to preserve my former student's anonymity. All other aspects of the story are true. I also consulted with a published profile, which I don't cite here to protect her anonymity.

49   **the brain regions:** Ethan Kross et al., "Coping with Emotions Past: The Neural Bases of Regulating Affect Associated with Negative Autobiographical Memories," *Biological Psychiatry* 65 (2009): 361–366; and Ayna Baladi Nejad, Philippe Fossati, and Cedric Lemogne, "Self-Referential Processing, Rumination, and Cortical Midline Structures in Major Depression," *Frontiers in Human Neuroscience* 7 (2013): 666.

49   *zoom out:* Ethan Kross and Özlem Ayduk, "Self-Distancing: Theory, Research, and Current Directions," in *Advances in Experimental Social Psychology,* eds. J. Olson and M. Zanna (Amsterdam: Elsevier, 2017), 81–136; and John P. Powers and Kevin S. LaBar, "Regulating Emo-

tion Through Distancing: A Taxonomy, Neurocognitive Model, and Supporting Meta-analysis," *Neuroscience and Biobehavioral Reviews* 96 (2019): 155–173.

49   **psychological immune system:** See Daniel T. Gilbert et al., "Immune Neglect: A Source of Durability Bias in Affective Forecasting," *Journal of Personality and Social Psychology* 75 (1998): 617–638, for an introduction to the concept of a psychological immune system.

50   **paradigm for studying self-control:** Walter Mischel, *The Marshmallow Test: Mastering Self-Control* (New York: Little, Brown, 2014); and Walter Mischel, Yuichi Shoda, and Monica Rodriguez, "Delay of Gratification in Children," *Science* 244 (1989): 933–938.

50   **battling inner-voice rumination:** Özlem Ayduk, Walter Mischel, and Geraldine Downey, "Attentional Mechanisms Linking Rejection to Hostile Reactivity: The Role of 'Hot' Versus 'Cool' Focus," *Psychological Science* 13 (2002): 443–448. Also see Cheryl L. Rusting and Susan Nolen-Hoeksema, "Regulating Responses to Anger: Effects of Rumination and Distraction on Angry Mood," *Journal of Personality and Social Psychology* 74 (1998): 790–803.

51   **The downside of this approach:** Ethan Kross and Özlem Ayduk, "Facilitating Adaptive Emotional Analysis: Distinguishing Distanced-Analysis of Depressive Experiences from Immersed-Analysis and Distraction," *Personality and Social Psychology Bulletin* 34 (2008): 924–938.

51   **tool that therapists should employ:** Aaron T. Beck, "Cognitive Therapy: Nature and Relation to Behavior Therapy," *Behavior Therapy* 1 (1970): 184–200. Also see Rick E. Ingram and Steven Hollon, "Cognitive Therapy for Depression from an Information Processing Perspective," in *Personality, Psychopathology, and Psychotherapy Series: Information Processing Approaches to Clinical Psychology,* ed. R. E. Ingram (San Diego: Academic Press, 1986), 259–281.

51   ***not* thinking about your problems:** For a classic review of research pointing to the harmful effects of avoidance, see Edna B. Foa and Michael J. Kozak, "Emotional Processing of Fear: Exposure to Corrective Information," *Psychological Bulletin* 99 (1986): 20–35. As I mention in the text, people can distance to achieve different goals (i.e., to avoid their emotions, to mindfully accept them, to approach and analyze them). Like a hammer that can be used to pound a nail into the wall or rip it out, distancing has multiple applications. And like any tool, whether it's helpful or harmful depends on how and why people use it. In the work reviewed in this section of the chapter, I focus on a context in which research indicates that distancing is helpful: to aid

people in their attempts to actively reflect on and make sense of their negative experiences. For a more detailed discussion of these issues, see the conclusion and Ethan Kross and Özlem Ayduk, "Self-Distancing: Theory, Research, and Current Directions."

51   **powerful optical device:** Georgia Nigro and Ulric Neisser, "Point of View in Personal Memories," *Cognitive Psychology* 15 (1983): 467–482; John A. Robinson and Karen L. Swanson, "Field and Observer Modes of Remembering," *Memory* 1 (1993): 169–184. People tend to recall intense negative experiences from a self-immersed/first person perspective: Arnaud D'Argembau, "Phenomenal Characteristics of Autobiographical Memories for Positive, Negative, and Neutral Events," *Applied Cognitive Psychology* 17 (2003): 281–294; and Heather K. McIsaac and Eric Eich, "Vantage Point in Episodic Memory," *Psychonomic Bulletin and Review* 9 (2002): 146–150. However, memories of trauma and self-conscious experiences are more likely to be recalled from a self-distanced/observer perspective: Lucy M. Kenny et al., "Distant Memories: A Prospective Study of Vantage Point of Trauma Memories," *Psychological Science* 20 (2009): 1049–1052; and Meredith E. Coles et al., "Effects of Varying Levels of Anxiety Within Social Situations: Relationship to Memory Perspective and Attributions in Social Phobia," *Behaviour Research and Therapy* 39 (2001): 651–665. For discussion of the implications of this distinction for emotion regulation, see Ethan Kross and Özlem Ayduk, "Self-Distancing: Theory, Research, and Current Directions."

52   **fly-on-the-wall perspective:** Ethan Kross, Özlem Ayduk, and Walter Mischel, "When Asking 'Why' Does Not Hurt: Distinguishing Rumination from Reflective Processing of Negative Emotions," *Psychological Science* 16 (2005): 709–715.

52   **differences in the verbal stream:** The examples of verbal streams that I cite were pulled from Ethan Kross and Özlem Ayduk, "Making Meaning out of Negative Experiences by Self-Distancing," *Current Directions in Psychological Science* 20 (2011): 187–191.

53   **response to stress:** Özlem Ayduk and Ethan Kross, "Enhancing the Pace of Recovery: Self-Distanced Analysis of Negative Experiences Reduces Blood Pressure Reactivity," *Psychological Science* 19 (2008): 229–231. Also see Rebecca F. Ray, Frank H. Wilhelm, and James J. Gross, "All in the Mind's Eye? Anger Rumination and Reappraisal," *Journal of Personality and Social Psychology* 94 (2008): 133–145.

53   **dampened emotional activity in the brain:** Brittany M. Christian et al., "When Imagining Yourself in Pain, Visual Perspective Matters:

The Neural and Behavioral Correlates of Simulated Sensory Experiences," *Journal of Cognitive Neuroscience* 27 (2015): 866–875.

53 **less hostility and aggression:** Dominik Mischkowski, Ethan Kross, and Brad Bushman, "Flies on the Wall Are Less Aggressive: Self-Distancing 'in the Heat of the Moment' Reduces Aggressive Thoughts, Angry Feelings, and Aggressive Behavior," *Journal of Experimental Social Psychology* 48 (2012): 1187–1191. Also see Tamara M. Pfeiler et al., "Adaptive Modes of Rumination: The Role of Subjective Anger," *Cognition and Emotion* 31 (2017): 580–589.

53 **people with depression:** Ethan Kross et al., " 'Asking Why' from a Distance: Its Cognitive and Emotional Consequences for People with Major Depressive Disorder," *Journal of Abnormal Psychology* 121 (2012): 559–569; Ethan Kross and Özlem Ayduk, "Boundary Conditions and Buffering Effects: Does Depressive Symptomology Moderate the Effectiveness of Distanced-Analysis for Facilitating Adaptive Self-Reflection?," *Journal of Research in Personality* 43 (2009): 923–927; Emma Travers-Hill et al., "Beneficial Effects of Training in Self-Distancing and Perspective Broadening for People with a History of Recurrent Depression," *Behaviour Research and Therapy* 95 (2017): 19–28. For a summary of research on the clinical implications of distancing and a discussion of how it operates under different conditions, see Ethan Kross and Özlem Ayduk, "Self-Distancing: Theory, Research, and Current Directions."

53 **highly anxious parents:** Louis A. Penner et al., "Self-Distancing Buffers High Trait Anxious Pediatric Cancer Caregivers Against Short- and Longer-Term Distress," *Clinical Psychological Science* 4 (2016): 629–640.

53 **Philippe Verduyn:** Philippe Verduyn et al., "The Relationship Between Self-Distancing and the Duration of Negative and Positive Emotional Experiences in Daily Life," *Emotion* 12 (2012): 1248–1263. For a conceptual replication of the finding demonstrating that distancing reduces positive affect, see June Gruber, Allison G. Harvey, and Sheri L. Johnson, "Reflective and Ruminative Processing of Positive Emotional Memories in Bipolar Disorder and Healthy Controls," *Behaviour Research and Therapy* 47 (2009): 697–704. For experimental data supporting the delayed benefits of distancing, see Kross and Ayduk, "Facilitating Adaptive Emotional Analysis."

54 **we are all prone:** Özlem Ayduk and Ethan Kross, "From a Distance: Implications of Spontaneous Self-Distancing for Adaptive Self-

Reflection," *Journal of Personality and Social Psychology* 98 (2010): 809–829.

54   **Researchers at Stanford:** Ray, Wilhelm, and Gross, "All in the Mind's Eye?"

54   **Across the Atlantic:** Patricia E. Schartau, Tim Dalgleish, and Barnaby D. Dunn, "Seeing the Bigger Picture: Training in Perspective Broadening Reduces Self-Reported Affect and Psychophysiological Response to Distressing Films and Autobiographical Memories," *Journal of Abnormal Psychology* 118 (2009): 15–27.

54   **shrinking the size of an image:** Joshua Ian Davis, James J. Gross, and Kevin N. Ochsner, "Psychological Distance and Emotional Experience: What You See Is What You Get," *Emotion* 11 (2011): 438–444.

55   **higher GPAs:** David S. Yeager et al., "Boring but Important: A Self-Transcendent Purpose for Learning Fosters Academic Self-Regulation," *Journal of Personality and Social Psychology* 107 (2014): 558–580.

55   **1010 B.C.E.:** John S. Knox, "Solomon," *Ancient History Encyclopedia,* Jan. 25, 2017, www.ancient.eu/solomon/.

55   **As the Bible tells us:** Robert Alter, *The Hebrew Bible: A Translation with Commentary* (New York: W. W. Norton, 2018).

56   **"Solomon's Paradox":** Igor Grossmann and Ethan Kross, "Exploring Solomon's Paradox: Self-Distancing Eliminates the Self-Other Asymmetry in Wise Reasoning About Close Relationships in Younger and Older Adults," *Psychological Science* 25 (2014): 1571–1580.

57   **Lincoln later reflected:** Doris Kearns Goodwin, *Team of Rivals* (New York: Simon & Schuster, 2005).

57   **what wisdom actually is:** Igor Grossmann, "Wisdom in Context," *Perspectives on Psychological Science* 12 (2017): 233–257.

57   **associate wisdom with advanced age:** Igor Grossmann et al., "Reasoning About Social Conflicts Improves into Old Age," *PNAS* 107 (2010): 7246–7250. Also see Darrell A. Worthy et al., "With Age Comes Wisdom: Decision Making in Younger and Older Adults," *Psychological Science* 22 (2011): 1375–1380.

58   **happening to someone else:** Grossmann and Kross, "Exploring Solomon's Paradox"; and Alex C. Huynh et al., "The Wisdom in Virtue: Pursuit of Virtue Predicts Wise Reasoning About Personal Conflicts," *Psychological Science* 28 (2017): 1848–1856.

58 **choose to do nothing:** This tendency is referred to as the omission bias. Ilana Ritov and Jonathan Baron, "Reluctance to Vaccinate: Omission Bias and Ambiguity," *Journal of Behavioral Decision Making* 3 (1990): 263–277.

58 **and this is a big but:** This study included three different conditions in which people were asked to make medical decisions for someone other than the self. Participants were randomly assigned to assume the role of a physician making a decision for a patient, a medical director setting treatment policy for all patients, or a parent making a decision for a child. Each of these "making a decision for someone else" conditions produced judgments that were equivalent to one another and superior compared with when participants decided for themselves. I averaged across the response rates for all three conditions for the purpose of text. Brian J. Zikmund-Fisher et al., "A Matter of Perspective: Choosing for Others Differs from Choosing for Yourself in Making Treatment Decisions," *Journal of General Internal Medicine* 21 (2006): 618–622.

58 **18 million:** Global Cancer Observatory, "Globocan 2018," International Agency for Research on Cancer, World Health Organization, 1, gco.iarc.fr/today/data/factsheets/cancers/39-All-cancers-fact-sheet.pdf.

59 **avoid an "inside view":** Daniel Kahneman, *Thinking, Fast and Slow* (New York: Farrar, Straus and Giroux, 2011).

59 **decision making more generally:** Qingzhou Sun et al., "Self-Distancing Reduces Probability-Weighting Biases," *Frontiers in Psychology* 9 (2018): 611.

59 **information overload:** Jun Fukukura, Melissa J. Ferguson, and Kentaro Fujita, "Psychological Distance Can Improve Decision Making Under Information Overload via Gist Memory," *Journal of Experimental Psychology: General* 142 (2013): 658–665.

59 **roll back "loss aversion":** Evan Polman, "Self-Other Decision Making and Loss Aversion," *Organizational Behavior and Human Decision Processes* 119 (2012): 141–150; Flavia Mengarelli et al., "Economic Decisions for Others: An Exception to Loss Aversion Law," *PLoS One* 9 (2014): e85042; and Ola Andersson et al., "Deciding for Others Reduces Loss Aversion," *Management Science* 62 (2014): 29–36.

59 **2008 U.S. presidential election:** Ethan Kross and Igor Grossmann, "Boosting Wisdom: Distance from the Self Enhances Wise Reason-

ing, Attitudes, and Behavior," *Journal of Experimental Psychology: General* 141 (2012): 43–48.

60   **eased the conflict:** Özlem Ayduk and Ethan Kross, "From a Distance: Implications of Spontaneous Self-Distancing for Adaptive Self-Reflection."

60   **buffered against romantic decline:** Eli J. Finkel et al., "A Brief Intervention to Promote Conflict Reappraisal Preserves Marital Quality over Time," *Psychological Science* 24 (2013): 1595–1601.

62   **creating positive personal narratives:** For review, see Dan P. McAdams and Kate C. McLean, "Narrative Identity," *Current Directions in Psychological Science* 22 (2013): 233–238.

63   **temporal distancing:** Emma Bruehlman-Senecal and Özlem Ayduk, "This Too Shall Pass: Temporal Distance and the Regulation of Emotional Distress," *Journal of Personality and Social Psychology* 108 (2015): 356–375. Also see Emma Bruehlman-Senecal, Özlem Ayduk, and Oliver P. John, "Taking the Long View: Implications of Individual Differences in Temporal Distancing for Affect, Stress Reactivity, and Well-Being," *Journal of Personality and Social Psychology* 111 (2016): 610–635; S. P. Ahmed, "Using Temporal Distancing to Regulate Emotion in Adolescence: Modulation by Reactive Aggression," *Cognition and Emotion* 32 (2018): 812–826; and Alex C. Huynh, Daniel Y. J. Yang, and Igor Grossmann, "The Value of Prospective Reasoning for Close Relationships," *Social Psychological and Personality Science* 7 (2016): 893–902.

64   **James Pennebaker:** For reviews, see James W. Pennebaker, "Writing About Emotional Experiences as a Therapeutic Process," *Psychological Science* 8 (1997): 162–166; James W. Pennebaker and Cindy K. Chung, "Expressive Writing: Connections to Physical and Mental Health," in *The Oxford Handbook of Health Psychology,* ed. H. S. Friedman (Oxford: Oxford University Press, 2011), 417–437; also see Eva-Maria Gortner, Stephanie S. Rude, and James W. Pennebaker, "Benefits of Expressive Writing in Lowering Rumination and Depressive Symptoms," *Behavior Therapy* 37 (2006): 292–303; Denise M. Sloan et al., "Expressive Writing Buffers Against Maladaptive Rumination," *Emotion* 8 (2008): 302–306; and Katherine M. Krpan et al., "An Everyday Activity as a Treatment for Depression: The Benefits of Expressive Writing for People Diagnosed with Major Depressive Disorder," *Journal of Affective Disorders* 150 (2013): 1148–1151.

64   **creates distance from our experience:** Jiyoung Park, Özlem Ayduk, and Ethan Kross, "Stepping Back to Move Forward: Expres-

sive Writing Promotes Self-Distancing," *Emotion* 16 (2016): 349–364. As Park and colleagues discuss, this doesn't mean that distance is the only factor explaining why expressive writing helps.

Chapter Four: When I Become You

68   **frequency illusion:** Also referred to as the "Baader-Meinhof phenomenon, Baader-Meinhof," *Oxford English Dictionary*, April 6, 2020, https://www.oed.com/view/Entry/250279.

69   **LeBron James:** Interview by Michael Wilbon. Henry Abbott, "LeBron James' Post-decision Interviews," ESPN, July 9, 2010, https://www.espn.com/blog/truehoop/post/_/id/17856/lebron-james-post-decision-interviews and Jim Gray, "LeBron James 'The Decision,'" ESPN, July 8, 2010, https://www.youtube.com/watch?v=bHSLw8DLm20.

69   **Malala Yousafzai:** Malala Yousafzai, interview by Jon Stewart, *The Daily Show with Jon Stewart*, Oct. 8, 2013.

70   **Jennifer Lawrence:** Brooks Barnes, "Jennifer Lawrence Has No Appetite for Playing Fame Games," *New York Times*, Sept. 9, 2015.

70   **Gallic Wars:** Julius Caesar, *Caesar's Gallic War: With an Introduction, Notes, and Vocabulary by Francis W. Kelsey*, 7th ed. (Boston: Allyn and Bacon, 1895).

70   *The Education of Henry Adams:* Henry Adams, *The Education of Henry Adams: An Autobiography* (Cambridge, MA: Massachusetts Historical Society, 1918).

71   **powerful techniques:** Sally Dickerson and Margaret E. Kemeny, "Acute Stressors and Cortisol Responses: A Theoretical Integration and Synthesis of Laboratory Research," *Psychological Bulletin* 130 (2004): 355–391.

71   **public speaking:** Ethan Kross et al., "Self-Talk as a Regulatory Mechanism: How You Do It Matters," *Journal of Personality and Social Psychology* 106 (2014): 304–324.

72   **marker of negative emotion:** For a historical review and meta-analysis, see Allison M. Tackman et al., "Depression, Negative Emotionality, and Self-Referential Language: A Multi-lab, Multi-measure, and Multi-language-task Research Synthesis," *Journal of Personality and Social Psychology* 116 (2019): 817–834; and To'Meisha Edwards and Nicholas S. Holtzman, "A Meta-Analysis of Correlations Be-

tween Depression and First-Person Singular Pronoun Use," *Journal of Research in Personality* 68 (2017): 63–68.

72   **For example:** The two studies I discuss in the text were published after our work on self-talk. As the papers cited in the previous endnote demonstrate, however, research stretching back multiple decades had already revealed a link between first-person singular pronoun usage and negative affect. I present these newer studies as evidence for that link because they represent particularly compelling evidence of the relationship. Tackman et al., "Depression, Negative Emotionality, and Self-Referential Language: A Multi-lab, Multi-measure, and Multi-language-task Research Synthesis"; and Johannes C. Eichstaedt et al., "Facebook Language Predicts Depression in Medical Records," *Proceedings of the National Academy of Sciences of the United States of America* 115 (2018): 11203–11208.

73   *distanced self-talk:* For reviews, see Ethan Kross and Özlem Ayduk, "Self-Distancing: Theory, Research, and Current Directions"; and Ariana Orvell et al., "Linguistic Shifts: A Relatively Effortless Route to Emotion Regulation?," *Current Directions in Psychological Science* 28 (2019): 567–573.

73   **third-person "he" or "she":** It's worth asking whether using "they" for those who identify as nonbinary would lead to a similar result. Although we have not tested this idea directly, theoretically we would expect this pronoun to serve the same distancing, emotion regulatory function.

73   **Other experiments:** Kross et al., "Self-Talk as a Regulatory Mechanism"; Sanda Dolcos and Dolores Albarracin, "The Inner Speech of Behavioral Regulation: Intentions and Task Performance Strengthen When You Talk to Yourself as a You," *European Journal of Social Psychology* 44 (2014): 636–642; and Grossmann and Kross, "Exploring Solomon's Paradox." For other domains in which distanced self-talk has revealed benefits, see Celina Furman, Ethan Kross, and Ashley Gearhardt, "Distanced Self-Talk Enhances Goal Pursuit to Eat Healthier," *Clinical Psychological Science* 8 (2020): 366–373; Ariana Orvell et al., "Does Distanced Self-Talk Facilitate Emotion Regulation Across a Range of Emotionally Intense Experiences?," *Clinical Psychological Science* (in press); and Jordan B. Leitner et al., "Self-Distancing Improves Interpersonal Perceptions and Behavior by Decreasing Medial Prefrontal Cortex Activity During the Provision of Criticism," *Social Cognitive and Affective Neuroscience* 12 (2017): 534–543.

74   **2014 Ebola crisis:** Ethan Kross et al., "Third-Person Self-Talk Reduces Ebola Worry and Risk Perception by Enhancing Rational Thinking," *Applied Psychology: Health and Well-Being* 9 (2017): 387–409.

74   **most chatter-provoking scenarios:** Aaron C. Weidman et al., "Punish or Protect: How Close Relationships Shape Responses to Moral Violations," *Personality and Social Psychology Bulletin* 46 (2019).

75   **"shifters" refer to words:** Orvell et al., "Linguistic Shifts"; and Roman Jakobson, *Shifters, Verbal Categories, and the Russian Verb* (Cambridge, MA: Harvard University, Russian Language Project, Department of Slavic Languages and Literatures, 1957). For discussion, see Orvell et al., "Linguistic Shifts."

76   **within milliseconds:** For discussion, see Orvell et al., "Linguistic Shifts."

76   **One tiny second:** Jason S. Moser et al., "Third-Person Self-Talk Facilitates Emotion Regulation Without Engaging Cognitive Control: Converging Evidence from ERP and fMRI," *Scientific Reports* 7 (2017): 1–9.

76   **overtaxed people's executive functions:** Ibid.

76   **Catch-22 of sorts:** Orvell et al., "Linguistic Shifts."

77   **typed to himself in 1979:** Robert Ito, "Fred Rogers's Life in 5 Artifacts," *New York Times*, June 5, 2018.

78   **think of it as a *challenge*:** Jim Blascovich and Joe Tomaka, "The Biopsychosocial Model of Arousal Regulation," *Advances in Experimental Social Psychology* 28 (1996): 1–51; and Richard S. Lazarus and Susan Folkman, *Stress, Appraisal, and Coping* (New York: Springer, 1984).

79   **Several studies back:** For review, see Jeremy P. Jamieson, Wendy Berry Mendes, and Matthew K. Nock, "Improving Acute Stress Responses: The Power of Reappraisal," *Current Directions in Psychological Science* 22 (2013): 51–56. Also see Adam L. Alter et al., "Rising to the Threat: Reducing Stereotype Threat by Reframing the Threat as a Challenge," *Journal of Experimental Social Psychology* 46 (2010): 155–171; and Alison Wood Brooks, "Get Excited: Reappraising Preperformance Anxiety as Excitement," *Journal of Experimental Psychology: General* 143 (2014): 1144–1158.

79   **Seventy-five percent:** Kross et al., "Self-Talk as a Regulatory Mechanism."

80   **see it in people's bodies:** Jim Blascovich and Joe Tomaka, "The Biopsychosocial Model of Arousal Regulation"; Mark D. Seery, "Challenge or Threat? Cardiovascular Indexes of Resilience and Vulnerability to Potential Stress in Humans," *Neuroscience and Biobehavioral Reviews* 35 (2011): 1603–1610.

80   **cardiovascular systems functioned:** Lindsey Streamer et al., "Not I, but She: The Beneficial Effects of Self-Distancing on Challenge/Threat Cardiovascular Responses," *Journal of Experimental Social Psychology* 70 (2017): 235–241.

81   **Batman Effect:** Rachel E. White et al., "The 'Batman Effect': Improving Perseverance in Young Children," *Child Development* 88 (2017): 1563–1571. Stephanie and her colleagues have examined the Batman Effect in additional contexts. In one direction, they've shown that this tool can promote executive functioning among five-year-olds: Rachel E. White and Stephanie M. Carlson, "What Would Batman Do? Self-Distancing Improves Executive Function in Young Children," *Developmental Science* 19 (2016): 419–426. In other work, they've shown that this tool is uniquely effective for young children and vulnerable children characterized by low levels of self-control when they work on frustrating tasks that have no solution: Amanda Grenell et al., "Individual Differences in the Effectiveness of Self-Distancing for Young Children's Emotion Regulation," *British Journal of Developmental Psychology* 37 (2019): 84–100.

81   **loss of a parent:** Julie B. Kaplow et al., "Out of the Mouths of Babes: Links Between Linguistic Structure of Loss Narratives and Psychosocial Functioning in Parentally Bereaved Children," *Journal of Traumatic Stress* 31 (2018): 342–351.

82   *normalizing* **experiences:** Robert L. Leahy, "Emotional Schema Therapy: A Bridge over Troubled Waters," in *Acceptance and Mindfulness in Cognitive Behavior Therapy: Understanding and Applying New Therapies,* ed. J. D. Herbert and E. M. Forman (Hoboken, NJ: John Wiley & Sons, 2011), 109–131; and Blake E. Ashforth and Glen E. Kreiner, "Normalizing Emotion in Organizations: Making the Extraordinary Seem Ordinary," *Human Resource Management Review* 12 (2002): 215–235.

83   **Sheryl Sandberg:** Sheryl Sandberg Facebook Post About Her Husband's Death, Facebook, June 3, 2015, www.facebook.com/sheryl /posts/10155617891025177:0. Also see Sheryl Sandberg in conversation with Oprah Winfrey, *Super Soul Sunday,* June 25, 2017, http://

www.oprah.com/own-super-soul-sunday/the-daily-habit-the
-helped-sheryl-sandberg-heal-after-tragedy-video.

83   **gaining helpful emotional distance:** Park, Ayduk, and Kross, "Stepping Back to Move Forward."

84   **"generic 'you' ":** Ariana Orvell, Ethan Kross, and Susan Gelman, "How 'You' Makes Meaning," *Science* 355 (2017): 1299–1302. Also see Ariana Orvell, Ethan Kross, and Susan Gelman, "Lessons Learned: Young Children's Use of Generic-You to Make Meaning from Negative Experiences," *Journal of Experimental Psychology: General* 148 (2019): 184–191.

84   **another type of linguistic hack:** Orvell et al., "Linguistic Shifts."

85   **asked to learn from their experience:** Orvell, Kross, and Gelman, "How 'You' Makes Meaning."

Chapter Five: The Power and Peril of Other People

87   **Then he opened fire on them again:** Steven Gray, "How the NIU Massacre Happened," *Time,* Feb. 16, 2008, content.time.com/time /nation/article/0,8599,1714069,00.html.

88   **Amanda Vicary and R. Chris Fraley:** Amanda M. Vicary and R. Chris Fraley, "Student Reactions to the Shootings at Virginia Tech and Northern Illinois University: Does Sharing Grief and Support over the Internet Affect Recovery?," *Personality and Social Psychology Bulletin* 36 (2010): 1555–1563; report of the February 14, 2008, shootings at Northern Illinois University, https://www.niu.edu /forward/_pdfs/archives/feb14report.pdf; Susan Saulny and Monica Davey, "Gunman Kills at Least 5 at U.S. College," *New York Times,* Feb. 15, 2008; and Cheryl Corley and Scott Simon, "NIU Students Grieve at Vigil," NPR, Feb. 16, 2008, https://www.npr.org /templates/story/story.php?storyId=19115808&t=1586343329323.

89   **one Virginia Tech student:** Vicary and Fraley, "Student Reactions to the Shootings at Virginia Tech and Northern Illinois University."

90   **September 11 attacks:** Mark D. Seery et al., "Expressing Thoughts and Feelings Following a Collective Trauma: Immediate Responses to 9/11 Predict Negative Outcomes in a National Sample," *Journal of Consulting and Clinical Psychology* 76 (2008): 657–667. The measure used to index expressing emotions following 9/11 consisted of an

open-ended prompt asking participants to share their thoughts about 9/11. The authors used this prompt as a proxy for assessing people's tendency to express emotions with others (pp. 663, 665). Critically, the authors demonstrate that people who completed the open-ended prompt also reported seeking out more emotional support and venting to others after the attacks (p. 664).

For additional resources indicating that expressing emotions is not always beneficial, see Richard McNally, Richard J. Bryant, and Anke Ehlers, "Does Early Psychological Intervention Promote Recovery from Posttraumatic Stress?," *Psychological Science in the Public Interest* 4 (2003): 45–79; Arnold A. P. van Emmerik et al., "Single Session Debriefing After Psychological Trauma: A Meta-analysis," *Lancet* 360 (2002): 766–771; George A. Bonanno, "Loss, Trauma, and Human Resilience: Have We Underestimated the Human Capacity to Thrive After Extremely Aversive Events?," *American Psychologist* 59 (2004): 20–28; Bushman, "Does Venting Anger Feed or Extinguish the Flame?"; Bushman et al., "Chewing on It Can Chew You Up"; and Rimé, "Emotion Elicits the Social Sharing of Emotion."

91   **earliest proponents of this approach:** Aristotle, *Poetics* (Newburyport, MA: Pullins, 2006). Also see, Brad J. Bushman, "Catharsis of Aggression," in *Encyclopedia of Social Psychology,* ed. Roy F. Baumeister and Kathleen D. Vohs (Thousand Oaks, CA: Sage, 2007), 135–137; and The Editors of *Encyclopaedia Britannica,* "Catharsis," *Encyclopaedia Britannica.*

91   **Sigmund Freud and his mentor:** Josef Breuer and Sigmund Freud, *Studies on Hysteria, 1893–1895* (London: Hogarth Press, 1955).

92   **earlier stage of our development:** I drew on Bernard Rimé's excellent synthesis of the role that developmental processes play in establishing emotion regulation as an interpersonal process for this section. Rimé, "Emotion Elicits the Social Sharing of Emotion."

93   **basic need we have to belong:** Roy F. Baumeister and Mark R. Leary, "The Need to Belong: Desire for Interpersonal Attachments as a Fundamental Human Motivation," *Psychological Bulletin* 117 (1995): 497–529.

93   **"tend and befriend" response:** Shelley E. Taylor, "Tend and Befriend: Biobehavioral Bases of Affiliation Under Stress," *Current Directions in Psychological Science* 15 (2006): 273–77.

93   **seek out other people:** Research indicates that simply thinking about caring for others, activating a mental snapshot of them, is sufficient for activating an inner coach like a script in people's heads.

According to psychologists Mario Mikulincer and Phillip Shaver, two pioneers in attachment research, the unspoken mental script goes like this: "If I encounter an obstacle and/or become distressed, I can approach a significant other for help; he or she is likely to be available and supportive; I will experience relief and comfort as a result of proximity to this person; I can then return to other activities." Mario Mikulincer et al., "What's Inside the Minds of Securely and Insecurely Attached People? The Secure-Base Script and Its Associations with Attachment-Style Dimensions," *Journal of Personality and Social Psychology* 97 (2002): 615–633.

   This script idea played into a set of studies I performed in 2015 with my colleague, Cornell psychologist Vivian Zayas, and her students, to examine whether glancing at pictures of attachment figures would have implications for helping people manage chatter. Specifically, we asked people to think about a negative experience that caused chatter and then asked people to look at a picture of either their mother or someone else's mother. Just as Mikulincer and Shaver would have predicted, looking at a picture of their mother reduced their emotional pain; they rated themselves as feeling much better. Emre Selcuk et al., "Mental Representations of Attachment Figures Facilitate Recovery Following Upsetting Autobiographical Memory Recall," *Journal of Personality and Social Psychology* 103 (2012): 362–378.

94   **emotional needs over our cognitive ones:** Christelle Duprez et al., "Motives for the Social Sharing of an Emotional Experience," *Journal of Social and Personal Relationships* 32 (2014): 757–787. Also see Lisanne S. Pauw et al., "Sense or Sensibility? Social Sharers' Evaluations of Socio-affective vs. Cognitive Support in Response to Negative Emotions," *Cognition and Emotion* 32 (2018): 1247–1264.

94   **interlocutors tend to miss these cues:** Lisanne S. Pauw et al., "I Hear You (Not): Sharers' Expressions and Listeners' Inferences of the Need for Support in Response to Negative Emotions," *Cognition and Emotion* 33 (2019): 1129–1243.

95   **co-rumination:** Amanda J. Rose, "Co-rumination in the Friendships of Girls and Boys," *Child Development* 73 (2002): 1830–1843; Jason S. Spendelow, Laura M. Simonds, and Rachel E. Avery, "The Relationship Between Co-rumination and Internalizing Problems: A Systematic Review and Meta-analysis," *Clinical Psychology and Psychotherapy* 24 (2017): 512–527; Lindsey B. Stone et al., "Co-rumination Predicts the Onset of Depressive Disorders During Adolescence,"

Journal of Abnormal Psychology 120 (2011): 752–757; and Benjamin L. Hankin, Lindsey Stone, and Patricia Ann Wright, "Co-rumination, Interpersonal Stress Generation, and Internalizing Symptoms: Accumulating Effects and Transactional Influences in a Multi-wave Study of Adolescents," *Developmental Psychopathology* 22 (2010): 217–235. Also see Rimé, "Emotion Elicits the Social Sharing of Emotion."

95   **when it comes to our inner voice:** For a discussion of the role that spreading activation theories play in rumination, see Rusting and Nolen-Hoeksema, "Regulating Responses to Anger."

97   **most effective verbal exchanges:** Andrew C. High and James Price Dillard, "A Review and Meta-analysis of Person-Centered Messages and Social Support Outcomes," *Communication Studies* 63 (2012): 99–118; Frederic Nils and Bernard Rimé, "Beyond the Myth of Venting: Social Sharing Modes Determine Emotional and Social Benefits from Distress Disclosure," *European Journal of Social Psychology* 42 (2012): 672–681; Stephen J. Lepore et al., "It's Not That Bad: Social Challenges to Emotional Disclosure Enhance Adjustment to Stress," *Anxiety, Stress, and Coping* 17 (2004): 341–361; Anika Batenburg and Enny Das, "An Experimental Study on the Effectiveness of Disclosing Stressful Life Events and Support Messages: When Cognitive Reappraisal Support Decreases Emotional Distress, and Emotional Support Is Like Saying Nothing at All," *PLoS One* 9 (2014): e114169; and Stephanie Tremmel and Sabine Sonnentag, "A Sorrow Halved? A Daily Diary Study on Talking About Experienced Workplace Incivility and Next-Morning Negative Affect," *Journal of Occupational Health Psychology* 23 (2018): 568–583.

97   **prefer to not cognitively reframe:** Gal Sheppes, "Transcending the 'Good and Bad' and 'Here and Now' in Emotion Regulation: Costs and Benefits of Strategies Across Regulatory Stages," *Advances in Experimental Social Psychology* 61 (2020). For further discussion of the role that time plays in social exchanges, see Rimé, "Emotion Elicits the Social Sharing of Emotion."

98   **Once the hostage takers understood:** Christopher S. Wren, "2 Give Up After Holding 42 Hostages in a Harlem Bank," *New York Times,* April 19, 1973; Barbara Gelb, "A Cool-Headed Cop Who Saves Hostages," *New York Times,* April 17, 1977; Gregory M. Vecchi et al., "Crisis (Hostage) Negotiation: Current Strategies and Issues in High-Risk Conflict Resolution," *Aggression and Violent Behavior* 10 (2005): 533–551; Gary Noesner, *Stalling for Time* (New York: Random House, 2010); "Police Negotiation Techniques from the NYPD

Crisis Negotiations Team," Harvard Law School, Nov. 11, 2019, https://www.pon.harvard.edu/daily/crisis-negotiations/crisis -negotiations-and-negotiation-skills-insights-from-the-new-york -city-police-department-hostage-negotiations-team/.

99   **diversify their sources of support:** Elaine O. Cheung, Wendi L. Gardner, and Jason F. Anderson, "Emotionships: Examining People's Emotion-Regulation Relationships and Their Consequences for Well-Being," *Social Psychological and Personality Science* 6 (2015): 407–414.

100   **global grassroots movement:** It Gets Better Project, itgetsbetter .org/; "How It All Got Started," https://itgetsbetter.org/blog /initiatives/how-it-all-got-started/; Brian Stelter, "Campaign Offers Help to Gay Youths," *New York Times,* Oct. 18, 2010; and Dan Savage, "Give 'Em Hope," *The Stranger,* Sept. 23, 2010.

101   **psychological debriefing:** McNally, Bryant, and Ehlers, "Does Early Psychological Intervention Promote Recovery from Posttraumatic Stress?"; and van Emmerik et al., "Single Session Debriefing After Psychological Trauma."

102   **powerful neurobiological experience:** For reviews of the empathy literature, see Zaki, *War for Kindness;* de Waal and Preston, "Mammalian Empathy"; and Erika Weisz and Jamil Zaki, "Motivated Empathy: A Social Neuroscience Perspective," *Current Opinion in Psychology* 24 (2018): 67–71.

103   **damages not only our self-esteem:** The relationship scientists Eshkol Rafaeli and Marci Gleason offer an incisive review of the social-support literature in Eshkol Rafaeli and Marci Gleason, "Skilled Support Within Intimate Relationships," *Journal of Family Theory and Review* 1 (2009): 20–37. They also provide a detailed discussion of the myriad additional ways that visible support can backfire. They note that it may focus attention on the source of stress, enhance how indebted one feels to a partner, highlight relationship inequities, and be perceived as hostile when the support is delivered with criticism (however well-intentioned).

103   **New York bar exam:** Niall Bolger, Adam Zuckerman, and Ronald C. Kessler, "Invisible Support and Adjustment to Stress," *Journal of Personality and Social Psychology* 79 (2000): 953–61. For an experimental conceptual replication of these results, see Niall Bolger and David Amarel, "Effects of Social Support Visibility on Adjustment to Stress: Experimental Evidence," *Journal of Personality and Social Psychology* 92 (2007): 458–475.

104   **A study on marriages:** Yuthika U. Girme et al., "Does Support Need to Be Seen? Daily Invisible Support Promotes Next Relationship Well-Being," *Journal of Family Psychology* 32 (2018): 882–893.

104   **meeting their self-improvement goals:** Yuthika U. Girme, Nickola C. Overall, and Jeffry A. Simpson, "When Visibility Matters: Short-Term Versus Long-Term Costs and Benefits of Visible and Invisible Support," *Personality and Social Psychology Bulletin* 39 (2013): 1441–1454.

104   **insights into the circumstances:** Katherine S. Zee and Niall Bolger, "Visible and Invisible Social Support: How, Why, and When," *Current Directions in Psychological Science* 28 (2019): 314–320. Also see Katherine S. Zee et al., "Motivation Moderates the Effects of Social Support Visibility," *Journal of Personality and Social Psychology* 114 (2018): 735–765.

105   **Caring physical contact:** Brittany K. Jakubiak and Brooke C. Feeney, "Affectionate Touch to Promote Relational, Psychological, and Physical Well-Being in Adulthood: A Theoretical Model and Review of the Research," *Personality and Social Psychology Review* 21 (2016): 228–252.

105   **one second of contact:** Sander L. Koole, Mandy Tjew A. Sin, and Iris K. Schneider, "Embodied Terror Management: Interpersonal Touch Alleviates Existential Concerns Among Individuals with Low Self-Esteem," *Psychological Science* 25 (2014): 30–37.

105   **teddy bear:** Ibid.; and Kenneth Tai, Xue Zheng, and Jayanth Narayanan, "Touching a Teddy Bear Mitigates Negative Effects of Social Exclusion to Increase Prosocial Behavior," *Social Psychological and Personality Science* 2 (2011): 618–626.

105   **result of the brain:** Francis McGlone, Johan Wessberg, and Hakan Olausson, "Discriminative and Affective Touch: Sensing and Feeling," *Neuron* 82 (2014): 737–751. For discussion of the role that C-fibers play in social support, see Jakubiak and Feeney, "Affectionate Touch to Promote Relational, Psychological, and Physical Well-Being in Adulthood."

105   **social organ:** India Morrison, Line S. Loken, and Hakan Olausson, "The Skin as a Social Organ," *Experimental Brain Research* 204 (2009): 305–314.

106   **nature of co-rumination via social media:** David S. Lee et al., "When Chatting About Negative Experiences Helps—and When It Hurts: Distinguishing Adaptive Versus Maladaptive Social Support in

Computer-Mediated Communication," *Emotion* 20 (2020): 368–375. For additional evidence indicating that the social sharing processes generalize to social media interactions, see Mina Choi and Catalina L. Toma, "Social Sharing Through Interpersonal Media."

Chapter Six: Outside In

108 **In 1963:** Erik Gellman, Robert Taylor Homes, Chicago Historical Society, http://www.encyclopedia.chicagohistory.org/pages/2478.html.

108 **Robert Taylor Homes:** Aaron Modica, "Robert R. Taylor Homes, Chicago, Illinois (1959–2005)," BlackPast, Dec. 19, 2009, blackpast .org/aah/robert-taylor-homes-chicago-illinois-1959-2005; D. Bradford Hunt, "What Went Wrong with Public Housing in Chicago? A History of the Robert Taylor Homes," *Journal of the Illinois State Historical Society* 94 (2001): 96–123; Hodding Carter, *Crisis on Federal Street,* PBS (1987).

109 **Ming Kuo:** Frances E. Kuo, "Coping with Poverty: Impacts of Environment and Attention in the Inner City," *Environment and Behavior* 33 (2001): 5–34.

109 **Roger Ulrich:** Roger S. Ulrich, "View Through a Window May Influence Recovery from Surgery," *Science* 224 (1984): 420–421.

111 **green revelations have followed:** For recent reviews of the link between nature exposure and health, see Gregory N. Bratman et al., "Nature and Mental Health: An Ecosystem Service Perspective," *Science Advances* 5 (2019): eaax0903; Roly Russell et al., "Humans and Nature: How Knowing and Experiencing Nature Affect Well-Being," *Annual Review of Environmental Resources* 38 (2013): 473–502; Ethan A. McMahan and David Estes, "The Effect of Contact with Natural Environments on Positive and Negative Affect: A Meta-analysis," *Journal of Positive Psychology* 10 (2015): 507–519; and Terry Hartig et al., "Nature and Health," *Annual Review of Public Health* 35 (2014): 207–228.

111 **ten thousand individuals in England:** Mathew P. White et al., "Would You Be Happier Living in a Greener Urban Area? A Fixed-Effects Analysis of Panel Data," *Psychological Science* 24 (2013): 920–928.

111 **seven years younger:** Omid Kardan et al., "Neighborhood Greenspace and Health in a Large Urban Center," *Scientific Reports* 5 (2015): 11610.

111   **entire population of England:** Richard Mitchell and Frank Pop-
       ham, "Effect of Exposure to Natural Environment on Health In-
       equalities: An Observational Population Study," *Lancet* 372 (2008):
       1655–1660. Also see David Rojas-Rueda et al., "Green Spaces and
       Mortality: A Systematic Review and Meta-analysis of Cohort Stud-
       ies," *Lancet Planet Health* 3 (2019): 469–477.

112   **Stephen and Rachel Kaplan:** Rachel Kaplan and Stephen Kaplan,
       The Experience of Nature: A Psychological Perspective (New York: Cam-
       bridge University Press, 1989). I also drew on this article to tell the
       Kaplans' story: Rebecca A. Clay, "Green Is Good for You," *Monitor
       on Psychology* 32 (2001): 40.

112   **William James:** William James, *Psychology: The Briefer Course* (New
       York: Holt, 1892).

113   **brain's limited resources:** For an excellent discussion of the distinc-
       tion between voluntary and involuntary attention as it relates to na-
       ture and attention restoration, see Stephen Kaplan and Marc G.
       Berman, "Directed Attention as a Common Resource for Executive
       Functioning and Self-Regulation," *Perspectives on Psychological Science* 5
       (2010): 43–57. Also see Timothy J. Buschman and Earl K. Miller,
       "Top-Down Versus Bottom-Up Control of Attention in the Prefron-
       tal and Posterior Parietal Cortices," *Science* 315 (2007): 1860–1862.

113   **One now classic study:** Marc G. Berman, John Jonides, and Stephen
       Kaplan, "The Cognitive Benefits of Interacting with Nature," *Psycho-
       logical Science* 19 (2008): 1207–1212. Also see Terry Hartig et al.,
       "Tracking Restoration in Natural and Urban Field Settings," *Journal
       of Environmental Psychology* 23 (2003): 109–123.

114   **clinically depressed participants:** Marc G. Berman et al., "Inter-
       acting with Nature Improves Cognition and Affect for Individuals
       with Depression," *Journal of Affective Disorders* 140 (2012): 300–305.

114   **Another satellite-imagery study:** Kristine Engemann et al., "Resi-
       dential Green Space in Childhood Is Associated with Lower Risk of
       Psychiatric Disorders from Adolescence into Adulthood," *Proceedings
       of the National Academy of Sciences of the United States of America* 116
       (2019): 5188–5193. Also see White et al., "Would You Be Happier
       Living in a Greener Urban Area?"

115   **Palo Alto, California:** Gregory N. Bratman et al., "Nature Experi-
       ence Reduces Rumination and Subgenual Prefrontal Cortex Activa-
       tion," *Proceedings of the National Academy of Sciences of the United States of
       America* 112 (2015): 8567–8572. For a conceptual replication at the
       behavioral level, see Gregory N. Bratman et al., "The Benefits of Na-

ture Experience: Improved Affect and Cognition," *Landscape and Urban Planning* 138 (2015): 41–50, which linked a nature (versus urban) walk with improved rumination, anxiety, positive affect, and working memory functioning.

115   **born and bred city dweller:** There's a natural level of skepticism that many people feel when they hear about these findings on the cognitive and emotional restorative effects of nature. Indeed, one clever set of studies found that people consistently underestimate how much interacting with green spaces will improve their mood. Elizabeth K. Nisbet and John M. Zelenski, "Underestimating Nearby Nature: Affective Forecasting Errors Obscure the Happy Path to Sustainability," *Psychological Science* 22 (2011): 1101–1106.

115   **68 percent of the world's population:** United Nations, Department of Economic and Social Affairs, Population Division, *World Urbanization Prospects: The 2018 Revision* (New York: United Nations, 2019); and Hannah Ritchie and Max Roser, "Urbanization," *Our World in Data* (2018, updated 2019), https://ourworldindata.org /urbanization#migration-to-towns-and-cities-is-very-recent-mostly -limited-to-the-past-200-years.

116   **six-minute video of neighborhood streets:** Bin Jiang et al., "A Dose-Response Curve Describing the Relationship Between Urban Tree Cover Density and Self-Reported Stress Recovery," *Environment and Behavior* 48 (2016): 607–629. Also see Daniel K. Brown, Jo L. Barton, and Valerie F. Gladwell, "Viewing Nature Scenes Positively Affects Recovery of Autonomic Function Following Acute-Mental Stress," *Environmental Science and Technology* 47 (2013): 5562–5569; Berman, Jonides, and Kaplan, "Cognitive Benefits of Interacting with Nature"; and McMahan and Estes, "Effect of Contact with Natural Environments on Positive and Negative Affect."

116   **improved performance on an attentional task:** Stephen C. Van Hedger et al., "Of Cricket Chirps and Car Horns: The Effect of Nature Sounds on Cognitive Performance," *Psychonomic Bulletin and Review* 26 (2019): 522–530.

116   **longer we're exposed:** Danielle F. Shanahan et al., "Health Benefits from Nature Experiences Depend on Dose," *Scientific Reports* 6 (2016): 28551. Also see Jiang et al., "Dose-Response Curve Describing the Relationship Between Urban Tree Cover Density and Self-Reported Stress Recovery."

117   **ReTUNE:** ReTUNE (Restoring Through Urban Nature Experience), The University of Chicago, https://appchallenge.uchicago

.edu/retune/, accessed March 4, 2020. ReTUNE app: https://retune -56d2e.firebaseapp.com/.

117   **Suzanne Bott:** Suzanne Bott, interview by Ethan Kross, Oct. 1, 2008.

118   **"the most dangerous place in Iraq":** Mark Kukis, "The Most Dangerous Place in Iraq," *Time,* Dec. 11, 2006.

119   **psychologist named Craig Anderson:** Craig L. Anderson, Maria Monroy, and Dacher Keltner, "Awe in Nature Heals: Evidence from Military Veterans, At-Risk Youth, and College Students," *Emotion* 18 (2018): 1195–1202.

120   **Awe is the wonder:** Jennifer E. Stellar et al., "Self-Transcendent Emotions and Their Social Functions: Compassion, Gratitude, and Awe Bind Us to Others Through Prosociality," *Emotion Review* 9 (2017): 200–207; Paul K. Piff et al., "Awe, the Small Self, and Prosocial Behavior," *Journal of Personality and Social Psychology* 108 (2015): 883–899; and Michelle N. Shiota, Dacher Keltner, and Amanda Mossman, "The Nature of Awe: Elicitors, Appraisals, and Effects on Self-Concept," *Cognition and Emotion* 21 (2007): 944–963.

120   **brain during awe-inspiring experiences:** Michiel van Elk et al., "The Neural Correlates of the Awe Experience: Reduced Default Mode Network Activity During Feelings of Awe," *Human Brain Mapping* 40 (2019): 3561–3574.

120   **brain responds when people meditate:** Judson A. Brewer et al., "Meditation Experience Is Associated with Differences in Default Mode Network Activity and Connectivity," *Proceedings of the National Academy of Sciences of the United States of America* 108 (2011): 20254–20259. For discussion of how the experience of awe relates to psychedelics in terms of underlying brain function, see van Elk et al., "The Neural Correlates of the Awe Experience: Reduced Default Mode Network Activity During Feelings of Awe." Also see Robin L. Carhart-Harris et al., "The Entropic Brain: A Theory of Conscious States Informed by Neuroimaging Research with Psychedelic Drugs," *Frontiers in Human Neuroscience* 3 (2014): 20.

120   **we developed this emotion:** For discussion, see Stellar et al., "Self-Transcendent Emotions and Their Social Functions."

121   **center of the world:** For example, see Yang Bai et al. "Awe, the Diminished Self, and Collective Engagement: Universals and Cultural Variations in the Small Self," *Journal of Personality and Social Psychology* 113 (2017): 185–209.

121 **synaptic flow of your thoughts:** van Elk et al., "Neural Correlates of the Awe Experience."

121 **similar ways as other distancing techniques:** For a similar argument, see Phuong Q. Le et al., "When a Small Self Means Manageable Obstacles: Spontaneous Self-Distancing Predicts Divergent Effects of Awe During a Subsequent Performance Stressor," *Journal of Experimental Social Psychology* 80 (2019): 59–66. This study also interestingly suggests that people who tend to spontaneously distance when reflecting on negative experiences may benefit the most from experiencing awe prior to delivering a stressful speech in terms of their cardiovascular stress response.

122 **purchasing a new watch:** Melanie Rudd, Kathleen D. Vohs, and Jennifer Aaker, "Awe Expands People's Perception of Time, Alters Decision Making, and Enhances Well-Being," *Psychological Science* 23 (2012): 1130–1136.

122 **linked with reduced inflammation:** Jennifer E. Stellar et al., "Positive Affect and Markers of Inflammation: Discrete Positive Emotions Predict Lower Levels of Inflammatory Cytokines," *Emotion* 15 (2015): 129–133.

122 **One set of studies:** Jennifer E. Stellar et al., "Awe and Humility," *Journal of Personality and Social Psychology* 114 (2018): 258–269.

122 **hallmark features of wisdom:** Grossmann and Kross, "Exploring Solomon's Paradox."

122 **caveat to consider:** Amie Gordon et al., "The Dark Side of the Sublime: Distinguishing a Threat-Based Variant of Awe," *Journal of Personality and Social Psychology* 113 (2016): 310–328.

124 **"What I battle hardest to do":** Rafael Nadal, *Rafa: My Story,* with John Carlin (New York: Hachette Books, 2013); Chris Chase, "The Definitive Guide to Rafael Nadal's 19 Bizarre Tennis Rituals," *USA Today,* June 5, 2019.

124 **compensatory control:** Mark J. Landau, Aaron C. Kay, and Jennifer A. Whitson, "Compensatory Control and the Appeal of a Structured World," *Psychological Bulletin* 141 (2015): 694–722.

124 **"It's a way of placing myself":** Nadal, *Rafa.*

124 **This might explain the global influence:** Maria Kondo, *The Life-Changing Magic of Tidying Up: The Japanese Art of Decluttering and Organizing* (Berkeley, CA: Ten Speed Press, 2014).

125 *perceptions of control:* As Mark Landau, Aaron Kay, and Jennifer Whitson deftly argue in their review, "Compensatory Control and

the Appeal of a Structured World," this topic has been the focus of a tremendous amount of research over the past sixty years and has been studied from a variety of perspectives.

125   **whether we try to achieve goals:** Albert Bandura, *Social Foundations of Thought and Action: A Social Cognitive Theory* (Englewood Cliffs, NJ: Prentice-Hall, 1986); and Bandura, *Self-Efficacy: The Exercise of Control* (New York: Freeman, 1997).

125   **improved physical health and emotional well-being:** For reviews, see Landau, Kay, and Whitson, "Compensatory Control and the Appeal of a Structured World"; D. H. Shapiro, Jr., C. E. Schwartz, and J. A. Astin, "Controlling Ourselves, Controlling Our World: Psychology's Role in Understanding Positive and Negative Consequences of Seeking and Gaining Control," *The American Psychologist* 51 (1996): 1213–1230; and Bandura, *Self-Efficacy: The Exercise of Control*. Also see Richard M. Ryan and Edward L. Deci, "Self-Determination Theory and the Facilitation of Intrinsic Motivation, Social Development, and Well-Being," *American Psychologist* 55 (2000): 68–78.

125   **heightened performance at school and work:** Michelle Richardson, Charles Abraham, and Rod Bond, "Psychological Correlates of University Students' Academic Performance: A Systematic Review and Meta-analysis," *Psychological Bulletin* 138 (2012): 353–387; Michael Schneider and Franzis Preckel, "Variables Associated with Achievement in Higher Education: A Systematic Review of Meta-analyses," *Psychological Bulletin* 143 (2017): 565–600; Alexander D. Stajkovic and Fred Luthans, "Self-Efficacy and Work-Related Performance: A Meta-analysis," *Psychological Bulletin* 124 (1998): 240–261.

125   **more satisfying interpersonal relationships:** Toni L. Bisconti and C. S. Bergeman, "Perceived Social Control as a Mediator of the Relationships Among Social Support, Psychological Well-Being, and Perceived Health," *Gerontologist* 39 (1999): 94–103; Tanya S. Martini, Joan E. Grusec, and Silvia C. Bernardini, "Effects of Interpersonal Control, Perspective Taking, and Attributions on Older Mothers' and Adult Daughters' Satisfaction with Their Helping Relationships," *Journal of Family Psychology* 15 (2004): 688–705.

125   **causes our chatter to spike:** For discussion, see Nolen-Hoeksema, Wisco, and Lyubomirsky, "Rethinking Rumination."

125   **propels us to try to regain it:** Another resource that people frequently utilize to enhance their sense of control is religion, which provides people with order, structure, and organization on practical

and spiritual levels. Aaron C. Kay et al., "God and the Government: Testing a Compensatory Control Mechanism for the Support of External Systems," *Journal of Personality and Social Psychology* 95 (2008): 18–35. For discussion, see Landau, Kay, and Whitson, "Compensatory Control and the Appeal of a Structured World."

125 **easier to navigate and more predictable:** Landau, Kay, and Whitson, "Compensatory Control and the Appeal of a Structured World."

125 **illusory patterns:** Jennifer A. Whitson and Adam D. Galinsky, "Lacking Control Increases Illusory Pattern Perception," *Science* 322 (2008): 115–117.

126 **the one with the structured border:** Keisha M. Cutright, "The Beauty of Boundaries: When and Why We Seek Structure in Consumption," *Journal of Consumer Research* 38 (2012): 775–790. Also see, Samantha J. Heintzelman, Jason Trent, and Laura A. King, "Encounters with Objective Coherence and the Experience of Meaning in Life," *Psychological Science* 24 (2013): 991–998.

126 **reading about the world:** Alexa M. Tullett, Aaron C. Kay, and Michael Inzlicht, "Randomness Increases Self-Reported Anxiety and Neurophysiological Correlates of Performance Monitoring," *Social Cognitive and Affective Neuroscience* 10 (2015): 628–635.

126 **perceive in their surroundings:** Catherine E. Ross, "Neighborhood Disadvantage and Adult Depression," *Journal of Health and Social Behavior* 41 (2000): 177–187.

126 **subset of people:** Not all people diagnosed with obsessive-compulsive disorder are motivated to establish order in their surroundings: Miguel Fullana, "Obsessions and Compulsions in the Community: Prevalence, Interference, Help-Seeking, Developmental Stability, and Co-occurring Psychiatric Conditions," *American Journal of Psychiatry* 166 (2009): 329–336.

127 **proliferation of conspiracy theories:** For discussion, see Landau, Kay, and Whitson, "Compensatory Control and the Appeal of a Structured World."

Chapter Seven: Mind Magic

130 **Franz Anton Mesmer:** I used the following resources to tell Mesmer's story: George J. Makari, "Franz Anton Mesmer and the Case of the Blind Pianist," *Hospital and Community Psychiatry* 45 (1994): 106–

110; Derek Forrest, "Mesmer," *International Journal of Clinical and Experimental Hypnosis* 50 (2001): 295–308; Douglas J. Lanska and Joseph T. Lanska, "Franz Anton Mesmer and the Rise and Fall of Animal Magnetism: Dramatic Cures, Controversy, and Ultimately a Triumph for the Scientific Method," in *Brain, Mind, and Medicine: Essays in Eighteenth-Century Neuroscience,* ed. Harry Whitaker (New York: Springer, 2007), 301–320; Sadie F. Dingfelder, "The First Modern Psychology Study: Or How Benjamin Franklin Unmasked a Fraud and Demonstrated the Power of the Mind," *Monitor on Psychology* 41 (2010), www.apa.org/monitor/2010/07-08/franklin; and David A. Gallo and Stanley Finger, "The Power of a Musical Instrument: Franklin, the Mozarts, Mesmer, and the Glass Armonica," *History of Psychology* 3 (2000): 326–343.

133   **didn't miss this point:** Benjamin Franklin, *Report of Dr. Benjamin Franklin, and Other Commissioners, Charged by the King of France, with the Examination of Animal Magnetism, as Now Practiced at Paris* (London: printed for J. Johnson, 1785).

133   **until the mid-twentieth century:** This dramatic jump forward is largely owed to an anesthesiologist named Henry Beecher, who published an article in 1955 called "The Powerful Placebo": Henry Beecher, "The Powerful Placebo," *Journal of the American Medical Association* 159 (1955): 1602–1606.

133   **ancient human tradition:** The Editors of *Encyclopaedia Britannica,* "Amulet," *Encyclopaedia Britannica.*

133   **mythical seal:** Joseph Jacobs and M. Seligsohn, "Solomon, Seal of," *Jewish Encyclopedia,* www.jewishencyclopedia.com/articles/13843 -solomon-seal-of.

133   **symbol of good fortune:** Mukti J. Campion, "How the World Loved the Swastika—Until Hitler Stole It," *BBC News,* Oct. 23, 2014, www.bbc.com/news/magazine-29644591.

134   **worry dolls:** Charles E. Schaefer and Donna Cangelosi, *Essential Play Therapy Techniques: Time-Tested Approaches* (New York: The Guilford Press, 2016).

134   **Heidi Klum:** Dan Snierson, "Heidi Klum Reveals Victoria's Secret," *Entertainment Weekly,* Nov. 21, 2003.

134   **Michael Jordan:** NBA.com Staff, "Legends Profile: Michael Jordan," NBA, www.nba.com/history/legends/profiles/michael-jordan.

134   **healing practice of crystals has become big business:** Rina Raphael, "Is There a Crystal Bubble? Inside the Billion-Dollar 'Healing' Gemstone Industry," *Fast Company,* May 5, 2017.

134   **it's quite rational:** For an excellent discussion of the psychological gymnastics that explain how rational individuals endorse superstitious beliefs, see Jane Risen, "Believing What We Do Not Believe: Acquiescence to Superstitious Beliefs and Other Powerful Intuitions," *Psychological Review* 123 (2016): 182–207.

134   **Study after study demonstrates:** Yoni K. Ashar, Luke J. Chang, and Tor D. Wager, "Brain Mechanisms of the Placebo Effect: An Affective Appraisal Account," *Annual Review of Clinical Psychology* 13 (2017): 73–98; Ted J. Kaptchuk and Franklin G. Miller, "Placebo Effects in Medicine," *New England Journal of Medicine* 373 (2015): 8–9; and Tor D. Wager and Lauren Y. Atlas, "The Neuroscience of Placebo Effects: Connecting Context, Learning and Health," *Nature Reviews Neuroscience* 16 (2015): 403–418.

134   **irritable bowel syndrome patients:** Ted J. Kaptchuk et al., "Components of Placebo Effect: Randomized Controlled Trial in Patients with Irritable Bowel Syndrome," *British Medical Journal* 336 (2008): 999–1003.

134   **migraine sufferers:** Karin Meissner et al., "Differential Effectiveness of Placebo Treatments: A Systematic Review of Migraine Prophylaxis," *JAMA Internal Medicine* 173 (2013): 1941–1951.

134   **improved respiratory symptoms for asthmatics:** Michael E. Wechsler et al., "Active Albuterol or Placebo, Sham Acupuncture, or No Intervention in Asthma," *New England Journal of Medicine* 365 (2011): 119–126.

134   **varies notably across diseases and patients:** For examples, see Andrew L. Geers et al., "Dispositional Optimism Predicts Placebo Analgesia," *The Journal of Pain* 11 (2010): 1165–1171; Marta Pecina et al., "Personality Trait Predictors of Placebo Analgesia and Neurobiological Correlates," *Neuropsychopharmacology* 38 (2013): 639–646.

134   **injected a promising new chemical treatment:** C. Warren Olanow et al., "Gene Delivery of Neurturin to Putamen and Substantia Nigra in Parkinson Disease: A Double-Blind, Randomized, Controlled Trial," *Annals of Neurology* 78 (2015): 248–257. For additional evidence that placebos benefit Parkinson's disease, see Raul de la Fuente-Fernandez et al., "Expectation and Dopamine Release: Mechanism of the Placebo Effect in Parkinson's Disease," *Science* 293 (2001): 1164–1166; Christopher G. Goetz, "Placebo Response in Parkinson's Disease: Comparisons Among 11 Trials Covering Medical and Surgical Interventions," *Movement Disorders* 23 (2008): 690–699; American Parkinson Disease Association, "The Placebo Effect in

Clinical Trials in Parkinson's Disease," March, 6, 2017, www.apda parkinson.org/article/the-placebo-effect-in-clinical-trials-in -parkinsons-disease/.

135   **after participants completed:** Leonie Koban et al., "Frontal-Brainstem Pathways Mediating Placebo Effects on Social Rejection," *Journal of Neuroscience* 37 (2017): 3621–3631.

136   **help people with chatter:** The flip side to the emotionally fortifying boost of placebos holds as well. In a phenomenon dubbed the "nocebo" effect, believing that a substance will harm you has also been shown to have that effect in some circumstances. Paul Enck, Fabrizio Benedetti, and Manfred Schedlowski, "New Insights into the Placebo and Nocebo Responses," *Neuron* 59 (2008): 195–206.

136   **depression and anxiety:** For review, see Ashar, Chang, and Wager, "Brain Mechanisms of the Placebo Effect."

137   **several months:** Arif Khan, Nick Redding, and Walter A. Brown, "The Persistence of the Placebo Response in Antidepressant Clinical Trials," *Journal of Psychiatric Research* 42 (2008): 791–796.

137   **Tig Notaro:** Stuart Heritage, "Tig Notaro and Her Jaw-Dropping Cancer Standup Routine," *Guardian,* Oct. 19, 2012; Andrew Marantz, "Good Evening. Hello. I Have Cancer," *New Yorker,* Oct. 5, 2012; Vanessa Grigoriadis, "Survival of the Funniest," *Vanity Fair,* Dec. 18, 2012; and Tig Notaro, *Live,* 2012.

138   **brain is a prediction machine:** Andy Clark, "Whatever Next? Predictive Brains, Situated Agents, and the Future of Cognitive Science," *Behavioral and Brain Sciences* 36 (2013): 181–204.

138   **generalizes to our internal experiences:** Irving Kirsch, "Response Expectancy and the Placebo Effect," *International Review of Neurobiology* 138 (2018): 81–93; and Christian Büchel et al., "Placebo Analgesia: A Predictive Coding Perspective," *Neuron* 81 (2014): 1223–1239.

139   **strengthen our beliefs:** For an excellent discussion of the role that preconscious and deliberative processes play in placebo effects, see Ashar, Chang, and Wager, "Brain Mechanisms of the Placebo Effect"; Donald D. Price, Damien G. Finniss, and Fabrizio Benedetti, "A Comprehensive Review of the Placebo Effect: Recent Advances and Current Thought," *Annual Review of Psychology* 59 (2008): 565–590; and Karin Meissner and Klaus Linde, "Are Blue Pills Better Than Green? How Treatment Features Modulate Placebo Effects," *International Review of Neurobiology* 139 (2018): 357–378; John D. Jennings et al., "Physicians' Attire Influences Patients' Perceptions in the Urban

Outpatient Surgery Setting," *Clinical Orthopaedics and Related Research* 474 (2016): 1908–1918.

139 **rodents and other animals respond to placebos:** As reviewed in Ashar, Chang, and Wager, "Brain Mechanisms of the Placebo Effect." Also see R. J. Herrnstein, "Placebo Effect in the Rat," *Science* 138 (1962): 677–678; and Jian-You Gou et al., "Placebo Analgesia Affects the Behavioral Despair Tests and Hormonal Secretions in Mice," *Psychopharmacology* 217 (2011): 83–90; and K. R. Munana, D. Zhang, and E. E. Patterson, "Placebo Effect in Canine Epilepsy Trials," *Journal of Veterinary Medicine* 24 (2010): 166–170.

140 **brain and spinal cord:** Tor D. Wager and Lauren Y. Atlas, "The Neuroscience of Placebo Effects."

140 **brain's pleasure circuitry:** Hilke Plassmann et al., "Marketing Actions Can Modulate Neural Representations of Experienced Pleasantness," *Proceedings of the National Academy of Sciences* 105 (2008): 1050–1054.

140 **hunger hormone ghrelin:** Alia J. Crum et al., "Mind over Milkshakes: Mindsets, Not Just Nutrients, Determine Ghrelin Response," *Health Psychology* 30 (2011): 424–429.

140 **stronger for psychological outcomes:** Ashar, Chang, and Wager, "Brain Mechanisms of the Placebo Effect."

140 **placebos can act as enhancers:** Slavenka Kam-Hansen et al., "Altered Placebo and Drug Labeling Changes the Outcome of Episodic Migraine Attacks," *Science Translational Medicine* 6 (2014): 218ra5.

141 **potent persuasive device:** For a classic reference, see Richard E. Petty and John T. Cacioppo, "The Elaboration Likelihood Model of Persuasion," *Advances in Experimental Social Psychology* 19 (1986): 123–205.

141 **Ted Kaptchuk and his team:** Ted J. Kaptchuk et al., "Placebos Without Deception: A Randomized Controlled Trial in Irritable Bowel Syndrome," *PLoS One* 5 (2010): e15591.

142 **our own experiment:** Darwin Guevarra et al., "Are They Real? Non-deceptive Placebos Lead to Robust Declines in a Neural Biomarker of Emotional Reactivity," *Nature Communications* (in press).

142 **nondeceptive placebos:** James E. G. Charlesworth et al., "Effects of Placebos Without Deception Compared with No Treatment: A Systematic Review and Meta-analysis," *Journal of Evidence-Based Medicine* 10 (2017): 97–107.

143   **Bronislaw Malinowski:** Raymond W. Firth, "Bronislaw Malinowski: Polish-Born British Anthropologist," *Encyclopaedia Britannica,* Feb. 2019; Katharine Fletcher, "Bronislaw Malinowski—LSE pioneer of Social Anthropology," June 13, 2017, LSE History, https://blogs.lse.ac.uk/lsehistory/2017/06/13/bronislaw-malinowski-lse-pioneer-of-social-anthropology/; Michael W. Young and Bronislaw Malinowski, *Malinowski's Kiriwina: Fieldwork Photography, 1915–1918* (Chicago: University of Chicago Press, 1998).

143   **betel nuts:** Cindy Sui and Anna Lacey, "Asia's Deadly Secret: The Scourge of the Betel Nut," *BBC News,* https://www.bbc.com/news/health-3192120; "Bronislaw Malinowski (1884–1942)," *Lapham's Quarterly,* www.laphamsquarterly.org/contributors/malinowski.

144   **"I kick thee down":** Bronislaw Malinowski, *Argonauts of the Western Pacific: An Account of Native Enterprise and Adventure in the Archipelagoes of Melanesian New Guinea* (Long Grove, IL: Waveland Press, 2010), loc. 5492–5493, Kindle; Bronislaw Malinowski, "Fishing in the Trobriand Islands," *Man* 18 (1918): 87–92; Bronislaw Malinowski, *Man, Science, Religion, and Other Essays* (Boston: Beacon Press, 1948).

144   **psychology of human beings:** I drew from this excellent review on the psychology of rituals for this section of the book: Nicholas M. Hobson et al., "The Psychology of Rituals: An Integrative Review and Process-Based Framework," *Personality and Social Psychology Review* 22 (2018): 260–284.

144   **West Point:** "10 Facts: The United States Military Academy at West Point," American Battlefield Trust, www.battlefields.org/learn/articles/10-facts-united-states-military-academy-west-point.

144   **business world as well:** Samantha McLaren, "A 'No Shoes' Policy and 4 Other Unique Traditions That Make These Company Cultures Stand Out," Linkedin Talent Blog, Nov. 12, 2018, business.linkedin.com/talent-solutions/blog/company-culture/2018/unique-traditions-that-make-these-company-cultures-stand-out.

145   **Wade Boggs:** George Gmelch, "Baseball Magic," in *Ritual and Belief,* ed. David Hicks (Plymouth, UK: AltaMira Press, 2010): 253–262; Jay Brennan, "Major League Baseball's Top Superstitions and Rituals," Bleacher Report, Oct. 3, 2017, bleacherreport.com/articles/375113-top-mlb-superstitions-and-rituals; and Matthew Hutson, "The Power of Rituals," *Boston Globe,* Aug. 18, 2016.

145   **Steve Jobs:** Steve Jobs, Commencement Address, Stanford University, June 12, 2005, *Stanford News,* June 14, 2005.

145   **Michael Norton and Francesca Gino:** Michael I. Norton and Francesca Gino, "Rituals Alleviate Grieving for Loved Ones, Lovers, and Lotteries," *Journal of Experimental Psychology: General* 143 (2014): 266–272.

145   **naturally turn:** Martin Lang et al., "Effects of Anxiety on Spontaneous Ritualized Behavior," *Current Biology* 25 (2015): 1892–1897; Giora Keinan, "Effects of Stress and Tolerance of Ambiguity on Magical Thinking," *Journal of Personality and Social Psychology* 67 (1994): 48–55; and Stanley J. Rachman and Ray J. Hodgson, *Obsessions and Compulsions* (Upper Saddle River, NJ: Prentice-Hall, 1980).

145   **recited psalms:** Richard Sosis and W. Penn Handwerker, "Psalms and Coping with Uncertainty: Religious Israeli Women's Responses to the 2006 Lebanon War," *American Anthropologist* 113 (2011): 40–55.

145   **reciting the rosary:** Matthew W. Anastasi and Andrew B. Newberg, "A Preliminary Study of the Acute Effects of Religious Ritual on Anxiety," *Journal of Alternative and Complementary Medicine* 14 (2008): 163–165.

145   **consume fewer calories:** Allen Ding Tian et al., "Enacting Rituals to Improve Self-Control," *Journal of Personality and Social Psychology* 114 (2018): 851–876.

146   **"Don't Stop Believin'":** Alison Wood Brooks et al., "Don't Stop Believing: Rituals Improve Performance by Decreasing Anxiety," *Organizational Behavior and Human Decision Processes* 13 (2016): 71–85. There is also evidence indicating that performing rituals reduces activation in brain systems that become active when people experience anxiety. Nicholas M. Hobson, Devin Bonk, and Michael Inzlicht, "Rituals Decrease the Neural Response to Performance Failure," *PeerJ* 5 (2017): e3363.

146   **aren't simply habits or routines:** Hobson et al., "Psychology of Rituals."

146   **Australian Olympic swimmer Stephanie Rice:** Gary Morley, "Rice's Rituals: The Golden Girl of Australian Swimming," CNN, June 28, 2012, www.cnn.com/2012/06/28/sport/olympics-2012-stephanie-rice-australia/index.html.

148   **ritualized cleaning behaviors:** Lang et al., "Effects of Anxiety on Spontaneous Ritualized Behavior."

148   **socially rejected by their peers:** Rachel E. Watson-Jones, Harvey Whitehouse, and Cristine H. Legare, "In-Group Ostracism Increases

High-Fidelity Imitation in Early Childhood," *Psychological Science* 27 (2016): 34–42.

148   **desired goals:** E. Tory Higgins, "Self-Discrepancy: A Theory Relating Self and Affect," *Psychological Review* 94 (1987): 319–340; and Charles S. Carver and Michael F. Scheier, "Control Theory: A Useful Conceptual Framework for Personality-Social, Clinical, and Health Psychology," *Psychological Bulletin* 92 (1982): 111–135. Also see Earl K. Miller and Jonathan D. Cohen, "An Integrative Theory of Prefrontal Cortex Function," *Annual Review of Neuroscience* 24 (2001): 167–202.

149   **karaoke study:** Brooks et al., "Don't Stop Believing."

Conclusion

153   **our species didn't evolve:** This is not to say that meditation and mindfulness aren't useful. Like the other techniques reviewed in this chapter, they are tools that are useful in some contexts. The broader point is that it is not useful (or feasible) to continually focus on the present, because succeeding often requires us to reflect on the future and past.

153   **useful in small doses:** Dacher Keltner and James J. Gross, "Functional Accounts of Emotions," *Cognition and Emotion* 13 (1999): 467–480; and Randolph M. Nesse, "Evolutionary Explanations of Emotions," *Human Nature* 1 (1989): 261–289.

153   **impossible for them to feel pain:** U.S. National Library of Medicine, "Congenital Insensitivity to Pain," National Institutes of Health, Dec. 10, 2019, ghr.nlm.nih.gov/condition/congenital-insensitivity-to-pain#genes.

155   **into a curriculum:** The curriculum for this project focuses broadly on teaching students how to control their emotions using several of the strategies reviewed in *Chatter,* along with other empirically supported tools.

155   **the pilot study:** This study took place during the winter of 2019 in a high school in the northeastern United States. Students were randomly assigned to the toolbox curriculum or a "control" curriculum that taught students about the science of learning. The curricula were co-created by scientists (Angela Duckworth, Daniel Willingham,

John Jonides, Ariana Orvell, Benjamin Katz, and myself) and teachers (Rhiannon Killian and Keith Desrosiers).

157 **different situations:** For a discussion of the importance of flexibly using different emotion-management strategies, see Cecilia Cheng, "Cognitive and Motivational Processes Underlying Coping Flexibility: A Dual-Process Model," *Journal of Personal and Social Psychology* 84 (2003): 425–438; and George A. Bonanno and Charles L. Burton, "Regulatory Flexibility: An Individual Differences Perspective on Coping and Emotion Regulation," *Perspectives on Psychological Science* 8 (2013): 591–612.

157 **when used interchangeably:** James J. Gross, "Emotion Regulation: Current Status and Future Prospects," *Psychological Inquiry* 26 (2015): 1–26; Ethan Kross, "Emotion Regulation Growth Points: Three More to Consider," *Psychological Inquiry* 26 (2015): 69–71.

Index

Ethan Kross, PhD, is one of the world's leading experts on controlling the conscious mind. An award-winning professor at the University of Michigan and Ross School of Business, he is the director of the Emotion & Self Control Laboratory. He has participated in policy discussion at the White House and has been interviewed about his work on *CBS Evening News, Good Morning America,* and *NPR Morning Edition.* His pioneering research has been featured in *The New York Times, The New Yorker, The Wall Street Journal, USA Today, New England Journal of Medicine,* and *Science.* He completed his BA at the University of Pennsylvania and his PhD at Columbia University. This is his first book.